Bonded Cement-Based Material Overlays for the Repair, the Lining or the Strengthening of Slabs or Pavements

RILEM STATE-OF-THE-ART REPORTS
Volume 3

For other titles published in this series, go to
www.springer.com/series/8780

Benoît Bissonnette • Luc Courard • David W. Fowler
Jean-Louis Granju
Editors

Bonded Cement-Based Material Overlays for the Repair, the Lining or the Strengthening of Slabs or Pavements

State-of-the-Art Report of the RILEM Technical Committee 193-RLS

 Springer

Editors
Benoît Bissonnette
Université Laval
Québec, QC, Canada
benoit.bissonnette@gci.ulaval.ca

Luc Courard
Université de Liège
Liège, Belgium
luc.courard@ulg.ac.be

David W. Fowler
University of Texas
Austin, TX, USA
dwf@mail.utexas.edu

Jean-Louis Granju
INSA de Toulouse
Toulouse, France
jeanlouis.granju@sfr.fr

ISBN 978-94-007-1238-6   e-ISBN 978-94-007-1239-3
DOI 10.1007/978-94-007-1239-3
Springer Dordrecht Heidelberg London New York

Library of Congress Control Number: 2011922910

Cover design: SPi Publisher Services

Printed on acid-free paper

Springer is part of Springer Science+Business Media (www.springer.com)

Contents

7   **Design** 141
     M. Treviño, J.-L. Granju, H. Beushausen, A. Chabot, H. Mihashi and
     J. Silfwerbrand

8   **Practice and Quality Assurance** 157
     M. Vaysburd, B. Bissonnette and R. Morin

TC 193-RLS – Technical Committee Members

TC 193-RLS: Bonded cement-based material overlays for the repair, the lining
or the strengthening of slabs or pavements

Réparation, re-surfaçage ou renforcement des dallages industriels
ou des chaussées par un rechargement adhérent à base cimentaire

Membership

Name	Affiliation	Country	Status
B. Bissonnette	CRIB – Université Laval (Québec, QC)	Canada	Chairman
H. Beuschausen	University of Cape Town (Cape Town)	South Africa	Member
A. Chabot	LCPC Nantes (Nantes)	France	Associate
L. Courard	Université de Liège (Liège)	Belgium	Secretary
L. Czarnecki	Warsaw University of Technology (Warsaw)	Poland	Member
E. Dénarié	EPFL Lausanne (Lausanne)	Switzerland	Member
R. Gagné	University of Sherbrooke (Sherbrooke, QC)	Canada	Associate
A. Garbacz	Warsaw University of Technology (Warsaw)	Poland	Member
J.-L. Granju	LMDC – INSA Toulouse (Toulouse)	France	Honorary chair
D.W. Fowler	ICAR – University of Texas (Austin, TX)	USA	Member
H. Mihashi	Tohoku University (Sendai)	Japan	Associate
R. Morin	Ville de Montréal (Montréal, QC)	Canada	Member
A. Müller	CRIB – Université Laval (Québec, QC)	Canada	Member
J. Silfwerbrand	Swedish C&C Research Institute (Stockholm)	Sweden	Member
M. Treviño	CTR – University of Texas (Austin, TX)	USA	Associate
A. Turatsinze	LMDC – INSA Toulouse (Toulouse)	France	Member
A.M. Vaysburd	Vaycon Consulting (Baltimore, MD)	USA	Member
R. Walter	Technical University of Denmark (Lyngby)	Denmark	Associate

Foreword

Efforts are being devoted by researchers, scientists and engineers around the world towards improving the durability of concrete repairs. The aim of RILEM Technical Committee TC 193-RLS was to collect the fundamental body of knowledge in the field of concrete overlays. As a result, this *State-of-the-Art* report covers the basics for the development of design rules adapted for the achievement of durable bonded overlaid structures, either for repair or strengthening.

The STAR document is intended to provide a wide and accurate survey of current knowledge and practice, such as to identify rationally what is really understood, what the shortcomings are and what is still needed in view of establishing a consistent and reliable overlay design procedure. It serves as a reference source with regard to the construction process, condition evaluation of the existing structure, bond achievement, debonding mechanisms, structural behaviour, design methods and material selection. It also contains up-to-date information on quality control and assurance, as well as on repair and maintenance of overlaid structures.

As a final task of the committee, practical recommendations for reliable and durable overlays will be issued subsequently.

On behalf of all TC members, we would like here to express our gratitude to Professor Jean-Louis Granju, who initiated the project, put together the committee and supervised the activities until his well-deserved retirement.

B. Bissonnette, Chair of RILEM TC 193-RLS
L. Courard, Secretary of RILEM TC 193-RLS

Chapter 1
Introduction

J.-L. Granju, B. Bissonnette and L. Courard

1.1 Introduction

Concrete has been used in construction for more than a century now and a number of existing structures exhibit distresses and/or lack of load-carrying capacity. Repair and strengthening of existing structures are in fact among the biggest challenges civil engineers will have to face in the years to come. Moreover, the number of concrete structures keeps growing, and therefore repair or retrofitting needs keep increasing. Present concerns of sustainable development emphasizing rehabilitation instead of new construction, acts as incentive to this trend.

Among different approaches that can be considered for concrete rehabilitation, bonded overlays are often the most economical alternative. Overlays are particularly suitable in the case of structures with large surface areas, where it can be either poured or sprayed. The following types of applications are of special concern: slabs on grade (for example, industrial floors), pavements, bridge decks, walls and tunnels. Toppings and linings are also relevant to the same type of problem.

The primary purpose of overlays is to extend the life of the candidate concrete structures, either by restoring a smooth sound surface and/or the original load-carrying capacity, or by improving the load-carrying capacity by a thickness increase. For a slab or pavement in good condition, a 25% increase in thickness can nearly double its stiffness in some cases, resulting in nearly a 50% reduction in flexural stress and a significant increase of its service life.

Other reasons for overlaying slabs or concrete pavements include:

J.-L. Granju
Laboratoire Matériaux et Durabilité des Constructions (LMDC), UPS-INSA, Toulouse, France

B. Bissonnette
Centre de Recherche sur les Infrastructures en Béton (CRIB), Université Laval, Québec (QC), Canada

L. Courard
GeMMe – Building Materials, ArGEnCo Department, University of Liège, Belgium

- matching the level of an adjacent slab or element;
- replacing deteriorated or contaminated concrete and reinstating the protection of the structure (especially its reinforcement);
- providing a more durable wearing surface;
- improving the frictional characteristics of the surface for pavements or bridge decks;
- restoring architectural features such as colour or texture.

The potential for overlay applications worldwide sums up to multi-millions of square meters. Nevertheless, the durability of bonded overlay systems still draw concerns in the technical community because of bond sustainability problems encountered in a number of cases. In such events, debonding is generally initiated at the boundaries, joints and cracks in the overlay, and can be aggravated by curling effects in the debonding area. That leads to new cracks and increased debonding, which accelerate the damage process and soon leads to the need of renewed repairs.

At this time, there is still no accepted design approach or methodology that can warrant the practitioner a successful outcome of the repair. The available information essentially consists of quite general recommendations relying on experience and very simplistic design considerations.

In order to progress through understanding and knowledge, the 193-RLS RILEM Technical Committee was created. Its goal was to set the basis for the development of relevant design rules and technical recommendations for the achievement of durable bonded overlays (for repair or strengthening).

The primary concern in the scope of work of this TC is the "bond" between the overlay and the repaired structure (the latter is often referred to as the substrate). Indeed, it is the foremost factor in a bonded overlay system, provided that a durable repair material is used and that the residual concrete base is sound, and its durability is synonymous of durability of the whole composite system. A rational and reliable design requires a thorough understanding of the bond and debonding mechanisms. The quality of the bond depends on both the substrate preparation and the overlay placement procedure. In this respect, the quality of workmanship is of major importance. Debonding generally results from a combination of internal stresses induced by the gravity loads and the differential drying and thermal length changes between the overlay and the substrate. Irrespective of the source of internal stresses, the same mechanism governs debonding initiation and propagation.

Two main overlay families must be distinguished: on the one hand, the bridge deck overlays and, on the other hand, the overlays performed on slabs on ground or on pavements.

This State-of-the-Art document is the major deliverable of 193-RLS TC. It is intended to provide a wide and accurate survey of current knowledge and practice, such as to identify rationally what is really understood, what the shortcomings are, and from that on, what information needs to be generated to progress towards a real overlay design process. A subsequent task of the members of this TC will be the writing of Recommendations for reliable and durable repairs, which will be published in a separate document.

In addition, the State-of-the-Art report highlights inconsistencies in the terminology and definitions found in the scientific documentation, depending on the country, the sector of the industry, etc. The Recommendations document will propose some clarifications, based upon the consensus reached among the TC members.

This book is divided into ten chapters:

Chapter 1: Introduction

Chapter 2: Overlay: Decision and Construction Process

This chapter first reports, through flow charts, a typical decision process methodology. Then a review of all relevant aspects of design and construction is proposed: material selection for the overlay, reinforcement considerations if applicable (rebars, fibre reinforcement), interface with the substrate, joints, environmental effects, curing of the overlay, and quality assurance/quality control. Throughout this review, distinctions are made between the main types of overlay applications (slabs or pavements and bridge decks).

Chapter 3: Condition Evaluation of the Existing Structure Prior to Overlay

The first part of this chapter presents a guide to assess the condition of the substrate and to determine whether a bonded overlay is a relevant solution for the structure under evaluation. The second part provides a survey and detailed description of available test methods for condition evaluation of concrete structures.

Chapter 4: Bond

This chapter first presents an exhaustive overview of all the factors affecting the bond (favorably or unfavorably), from surface preparation to the use of bonding agents. The available testing procedures for bond characterization are then discussed. Finally, the existing recommendations to yield a durable bond are summarized.

Chapter 5: Structural Behavior

This chapter deals with the structural behavior of a composite system made of an existing substrate and a bonded cement-based overlay. Through a review of analytical models, the mechanical response of the repair composite is addressed, taking into account gravity loads, thermal and hygrometric strains, material ageing, as well as boundary conditions such as the degree of restraint and the presence of joints. Special attention is paid to restrained deformations, inherent to repaired concrete members. Then, an overview of relevant experimental work and numerical modeling is presented.

Chapter 6: Debonding

In this chapter, emphasis is put on debonding, which can lead to the failure of an overlaid system through partial or total loss of composite action. A summary of field observations and a survey of existing test methods to detect debonding are first proposed. Debonding mechanisms are then described, from initiation to propagation, as a result of length changes and/or repeated gravity loads (fatigue). Further on, proposed modeling approaches are discussed. Supported by experimental data, modeling is used to demonstrate and help understanding the benefits of different reinforcement/anchoring solutions. Finally, the influence of joints in an overlaid structure is reviewed.

Chapter 7: Design

This chapter summarizes the existing design rules, recommendations and requirements for bonded concrete overlays, as found in the guidelines of some of the foremost organizations around the world. The proposed approaches go from very simplistic criteria, such as minimum bond strength or stress level limitations, to slightly more sophisticated design methods.

Chapter 8: Practice and Quality Assurance

This chapter sums up the fundamentals of current good field practice, from substrate preparation to placement and curing of the overlay. It is completed with recommendations for proper quality control and quality assurance.

Chapter 9: Maintenance and Repair of Overlays

Although bonded overlays are an attractive and improving repair alternative, they still carry some uncertainty and, sometimes, unsuccessful repairs have to be dealt with. This chapter describes the suitable repair methods or techniques.

Chapter 10: Conclusion

Chapter 2
Overlay Design Process

D.W. Fowler and M. Treviño

Abstract Bonded concrete overlays have been used for nearly 100 year to extend the life of pavements, concrete slabs, bridge decks or other structural slabs. First, this chapter describes material selection, for slabs on grade and pavements; joints; and construction procedures including steel placement, environmental effects, and curing. Secondly, the bonded concrete overlay (BCO) process is described: the steps required in project selection; design of the BCO; construction, and quality assurance with flow charts included to provide a graphical overview.

2.1 Purpose of Overlays

Bonded concrete overlays (BCOs) have been used since 1909. The primary purpose of overlays is to extend the life of a concrete slab or pavement, bridge deck or other structural slab. It has been shown [1, 2] that as the remaining life of a pavement decreases due to distress, e.g. cracking, spalling or punchouts, the life can be extended significantly by the use of a bonded concrete overlay. For a slab or pavement in good condition, a 25% increase in thickness can nearly double the stiffness, resulting in nearly a 50% reduction in flexural stress.

Other reasons for overlaying concrete pavements or slabs include:

- provide an improved frictional surface for pavements or bridges;
- provide a smooth surface for industrial floors or buildings;
- increase the elevation of the top surface to match an adjacent slab;
- provide a more durable wearing surface;

D.W. Fowler
The University of Texas at Austin (TX), U.S.A.

M. Treviño
Center for Transportation Research, The University of Texas at Austin, U.S.A.

5

- repair corrosion-damaged slabs or bridge decks with sound, durable concrete; and
- provide architectural features such as color or texture.

2.2 Materials Selection

2.2.1 Slabs on Grade/Pavements

2.2.1.1 Concrete

The cement content of the BCO concrete must be high enough to ensure that the available paste is sufficient to achieve bond at the interface, which eliminates the need for a bonding agent, and to meet strength and permeability requirements. Reduced paste requirements may be met by using well-graded aggregates. Some specifications mandate minimum cement content, e.g. 390 kg/m^3 for normal overlay concrete and 490 kg/m^3 for dense overlay concrete. Lower levels of cement are desirable to reduce cost, reduce heat of hydration and reduce shrinkage. Reduced cost and reduced heat of hydration may also be accomplished by using fly ash as cement replacement; the addition of fly ash also has the advantages of improving durability and, in many cases, ultimate strength.

The water-cement ratio is determined by strength and durability (permeability) requirements. Generally, a water-cement ratio of 0.40 will provide good durability and strength. One specification requires a water-cement ratio of 0.40 for normal overlay concrete and 0.32 for dense overlay concrete.

Aggregates should be selected for workability; aggregates should also have adequate durability for the intended application. The paste requirement may be reduced by improving overall aggregate gradation as suggested by Shilstone [3, 4] and Crouch [5]. Shilstone suggests incorporating an intermediate aggregate, and this may be particularly helpful when using steel fibers.

Admixtures will often include air entraining agent, which will help workability and improve freezing and thawing resistance. The use of fly ash may require greater dosages of air entraining agent to achieve the same percentage of entrained air. High range water reducers are often specified, and the amount should be based on trial batches. The addition or retarders in hot weather helps to preserve workability until the concrete can be placed and finished without affecting strength development.

2.2.1.2 Reinforcement

Steel reinforcing and steel and synthetic fibers have been used as reinforcement. Steel reinforcing has been used in the form of tied bars and welded mats.

Steel fibers have been used successfully in overlays to control cracks and to minimize drying shrinkage cracking. Normal steel fiber contents for steel fiber reinforced concrete are normally in the range of 1 to 2% by volume. The amount of fibers required to achieve the desired result will depend on the type of fiber based on its bond characteristics.

Synthetic fibers should be used at a minimum of 0.25% by volume; however, significant benefits have been obtained by using up to 1% volume.

2.2.2 Structural Slabs and Decks

2.2.2.1 Concrete

Overlays for structural slabs and decks are generally used to replace removed concrete since the slab or deck cannot tolerate additional dead load and still continue to carry the design live load. One of the main applications of overlays for structural slabs is to replace deteriorated concrete due to corrosion of the steel reinforcing. The original concrete is removed to a depth of about 25 mm below the steel, and the overlay is used to reinstate the original surface. Generally, very durable, low permeability concrete is specified. Guidelines similar to those outlined for pavements are appropriate for structural slabs. In addition, latex, e.g. styrene-butadiene, is often added (15% latex solids by weight of cement) to provide improved bonding, greater flexural strength and decreased permeability.

2.2.2.2 Reinforcement

Steel bars are normally used for structural slabs and decks. Often the original bars are adequate, although if significant reduction in cross section due to corrosion has occurred, the affected bars should be spliced with adequate anchorage length or replaced. In some cases when extensive replacement of reinforcing is required, corrosion resistant bars, e.g. epoxy coated steel, galvanized, or stainless steel may be considered for use in highly corrosive environments.

Fibers, steel or synthetic, may be used for crack control and to minimize shrinkage cracking.

2.3 Joints

2.3.1 Slabs on Grade/Pavements

Jointed slabs and pavements usually require that the joints in the original slab or pavement be reinstated which is time consuming and costly. Sawing the joints and applying suitable high elongation joint filler is one possible solution. In some cases it may be possible to clean the joints of debris and apply joint filler. If it essential to reinstate the dowels at joints to prevent faulting, precast joint units have been used by removing the concrete for approximately 600 mm on each side of the joint and installing the units which have dowels in place.

The overlays must be constructed to maintain the integrity of the joints. Saw cutting followed by application of suitable high elongation joint filler may be the optimum solution.

2.3.2 Structural Slabs and Decks

Joints in structural slabs and decks are expansion joints. In some cases the nosings at the joints of the original concrete have spalled and deteriorated and must be repaired. The concrete in the overlay at the joints must have the durability and toughness to resist the impact of vehicular traffic, where applicable, and one solution is to use very strong, impact-resistant concrete headers on either side of the joints to which the overlay can be butted.

2.4 Construction Procedures

2.4.1 Steel Placement

Research has shown that the reinforcing placed on the substrate of a 75-mm thick slab with a 75-mm overlay cast on the surface will achieve the same pull-out bond strength as reinforcing placed at mid-depth of a 150-mm thick slab; all bars failed in tension [6]. Placing the reinforcing on the substrate saves construction time and labor.

2.4.2 Environmental Effects

Weather conditions during construction of the overlay can be critical. Hot, dry climates cause the most problems because of the excessive evaporation of water from the concrete that often results. High evaporation rates during placements can result in plastic shrinkage cracking. Evaporation rate is a function of wind velocity, relative humidity, concrete temperature, and air temperature. An evaporation rate of 1 kg/m^2 or higher has been suggested as sufficiently high to cause plastic shrinkage cracking, but even lower values may cause problems. The evaporation rate can be estimated from published monographs [7, 8] or calculated from the equation.

The evaporation rate should be monitored through construction by the use of a weather station which measures the four factors that influence evaporation. If the threshold value of evaporation rate is approached, action should be taken: discontinue placement of the overlay or provide measures to reduce evaporation including improved curing methods.

Concrete temperature should be monitored throughout construction. Temperature at placement has been shown to have a significant effect on performance [9]. Pavements placed in high temperatures in the summer have been found to have poor performance based on such indicators as crack spacing and distress occurrence compared to pavements placed in cooler weather. Therefore, for best performance the BCO placement temperature should be kept relatively low to avoid high temperatures of hydration. The surface of the substrate should not be permitted to exceed a temperature of 50°C [9].

Another environmental issue that can be detrimental to the concrete performance is the temperature differential that occurs in the hours following placement. High ambient temperature differentials within 24 hours after placement may cause extensive thermal cracking; drops in temperature from the peak high temperature to the low temperature should not exceed 15°C [9]. The largest drops in temperature in the concrete are usually associated with concrete placements in the morning in hot weather, which leads to the maximum hydration temperature occurring in combination with the maximum ambient temperature in the afternoon, followed by a large reduction in ambient temperature during the night. The rapid cooling of the surface can lead to thermal contraction strains that exceed the capacity of the concrete at that young age. Weather forecasts should be consulted to determine if placement conditions will be likely to cause problems.

If any of the adverse environmental conditions described in this section occurs during the placement of the concrete, placement should be avoided unless the conditions can be offset by such measures as:

- cooling the aggregates or concrete;
- special curing methods discussed in a following section; or
- use of fly ash as cement replacement to lower the heat of hydration.

2.4.3 Curing

Proper curing procedures are necessary to avoid excessive moisture loss at early ages that can result in plastic shrinkage and loss in tensile strength capacity at the

Table 2.1 Recommended curing for bonded concrete overlays

Condition	Recommendation
Evaporation below 0.5 kg/m²/hr	Membrane curing
Evaporation above 0.5 kg/m²/hr but below 1 kg/m²/hr	Membrane curing, plus evaporation retardant or fogging or wet mats, in place for 12 hours
Evaporation over 1 kg/m²/hr	Membrane curing, plus wet mat curing or fogging or other approved methods, in place 36 hours
Temperature drop in next 24 hours less than 15°C below temperature at time of paving	Membrane curing
Temperature drop in next 24 hours more than 15°C below temperature at time of paving	Membrane curing plus wet mats for 36 hours, or other approved methods

surface. Curing should begin as soon after placement and finishing as possible to minimize loss of bleed water. Curing may be accomplished by:

1. Application of curing compound.
   For textured or tined surfaces the spray application should be applied from two directions to ensure that the entire surface is coated.
2. Application of membrane curing.
   Various liquid sealing compounds, e.g. bituminous and paraffinic emulsions, coal tar cut-backs, pigmented and non-pigmented resin suspensions, or suspensions of wax or non-liquid protective coating such as sheet plastics or waterproof paper, are used to restrict evaporation of water.
3. Curing blankets.
   A covering of sacks, mattings, burlap, straw, or other suitable paper is placed over the surface to reduce evaporation and to reduce the temperature reduction at the surface. When used to reduce evaporation the blankets are generally wetted.
4. Monomolecular film.
   Monomolecular films (MMF) are compounds that form a thin monomolecular film to reduce rapid moisture loss from the concrete surface prior to curing. Another curing method should be used after the evaporation retardant is sprayed on. Research has shown, however, that the use of MMF followed by application of curing compound does not consistently provide less evaporation than curing compound alone.

Table 2.1 summarizes recommended curing procedures for bonded concrete overlays.

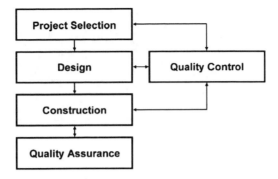

Fig. 2.1 BCO development process

2.5 The BCO Process

The BCO process is summarized in the flow chart in Figure 2.1. Each step of the process will be discussed.

2.5.1 Project Selection

The project selection involves several decisions as shown in the flow chart in Figure 2.2.

1. Need for rehabilitation, which is based on level of deterioration, age and increase in traffic loadings.
2. Availability of resources to the owner/agency with the responsibility for maintaining the highway.
3. Type of rehabilitation: overlay versus non-overlay.
4. Type of overlay if an overlay is selected. The overlay choices are portland cement concrete (PCC) and asphalt concrete. For PCC overlays the choice is between bonded and unbonded.
5. Timing and condition. A PCC overlay should be applied after structural failures have occurred but prior to functional failures at which time the pavement has reached a minimal level of serviceability. Figure 2.3 shows the pavement serviceability index (PSI) as a function of time and a hypothetical time at which structural failure and functional failure occurs.

Once the decision is made to construct a BCO, the next step is the design.

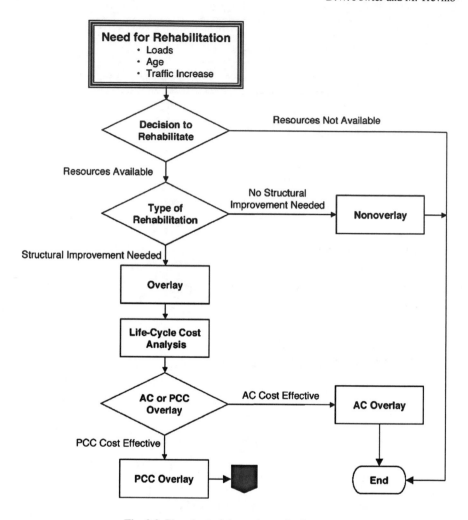

Fig. 2.2 Flowchart of the project selection stage

2.5.2 Design

The design decisions that must be made are:

- design period (overlay design life);
- traffic analysis; and
- remaining life of the original pavement.

The original pavement must be characterized, which usually requires:

- coring to determine modulus of elasticity, tensile strength, and thickness; and
- deflection testing.

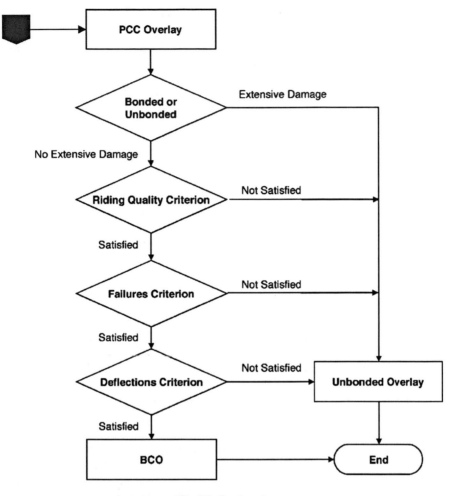

Fig. 2.2 Continued.

The thickness design is a function of:

- structural capacity of existing pavement; and
- structural capacity of BCO to fulfill future traffic requirements.

There are several design procedures available for performing the design, but a discussion of the methods is beyond the scope of this chapter.

2.5.3 Construction

Some of the construction-related considerations are:

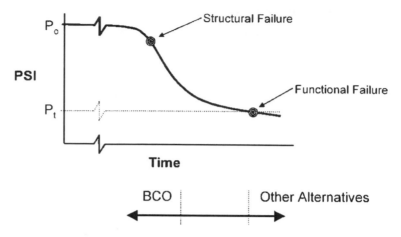

Fig. 2.3 Manifestation of structural and functional failure along the PSI curve

1. Materials and aggregates.
   Compatibility between the overlay and original pavement concrete is important, and the coefficient of thermal expansion and modulus of elasticity should be lower than the original concrete if possible. It has been found that aggregates used in the overlay should have lower thermal coefficients that can normally be obtained by using limestone materials.
2. Repairs.
   The original pavement must be repaired prior to overlay; the overlay will have no more integrity than the pavement to which it is bonded.
3. Surface preparation.
   The surface must be adequately cleaned, preferably by shot blasting using steel shot to a moderate roughness. The surface must be kept clean prior to placement of overlay concrete. Bonding agents are not necessary and should be avoided since they provide an additional step that can cause failure, e.g., a bonding agent that is allowed to cure prior to concrete placement, becoming a bond breaker [10].
4. Tied PCC shoulders.
   Tied PCC shoulders to the BCO provide slab edge support that in turn results in a reduction of stresses and deflections in the BCO and shoulder due to enhanced load transfer.
5. Critical weather conditions.
   Several research studies have shown that the following weather conditions during BCO construction may have adverse effects on the overlay performance [1, 9, 11, 12]:

   (a) It has generally been accepted that evaporation rates of ~ 1 kg/m^2/h or more can lead to plastic shrinkage cracking which may lead to greater probability of delamination. It should be noted, however, that with the wide range of chemical and mineral admixtures currently being used the critical evaporation rate to initiate plastic shrinkage cracking might be much lower than 1.0 kg/m^2. Some transportation agencies now require weather stations on paving construction projects to monitor evaporation rate.

(b) Substrate surface temperatures >50°C at concrete placement may lead to inadequate bonding.

(c) Daily temperature differentials of >15°C (between the high ambient temperature on the day of placement and the low night temperature following) may lead to poor bonding.

6. Special.
When evaporation rates approach or exceed the recommended maximum values, special curing, e.g., evaporation retardants, wet mats or fogging, should be required.

2.5.4 Quality Control/Quality Assurance (QA/QC)

A comprehensive QA/QC program is required to insure that the BCO provides the owner with a cost-effective pavement and results in a uniform, durable, safe, and low maintenance riding surface for the users. The program should include some or all of the following items:

- statistical sampling – samples from each subset of paving should be required;
- weather monitoring as described in the previous section relating to construction;
- materials tests;
- condition surveys of the completed BCO which includes sounding (to locate possible delamination) and mapping of cracks and other visible defects;
- pull-off tests (to determine bond strength);
- core tests; and
- deflection testing, e.g., falling weight deflectometer.

2.6 Conclusions

Bonded concrete overlays have many uses for highways, bridges and floors. There are many factors that must be taken into account in their design: material selection, joints, and construction procedures.

The use of bonded concrete overlays should be determined based on a rational project selection process, which considers the need for rehabilitation, availability of resources, type of rehabilitation, type of overlay and appropriateness of a BCO. The design should be conducted using a rational procedure to achieve the design life required by the owner. Construction considerations should include materials to insure compatibility of new and old concrete, repair of existing pavement, surface preparation, surface cleaning, weather conditions and special curing. A QA/QC program should be required to insure the quality of the finished BCO.

References

1. Delatte, N.J., Grater, S.F., Trevino-Frias, M, Fowler, D.W., and McCullough, B.F., Partial Construction Report of a Bonded Concrete Overlay on IH-10, El Paso, and Guide for Expedited BCO Design and Construction, Research Report 2911-5F, Center for Transportation Research, The University of Texas at Austin, 1996.
2. Van Metzinger, W., Lundy, J.R., McCullough, B.F., and Fowler, D.W., Research Report 1205-4F Design and Construction of Bonded Concrete Overlays, The University of Texas at Austin Center for Transportation Research, Austin, TX, January 1991.
3. Shilstone, J.M., Sr., Concrete Mixture Optimization, Concrete International, American Concrete Institute, Farmington Hills, Michigan, pp. 33–39, June 1990.
4. Shilstone, J., Sr., and Shilstone, J., Jr., Performance-Based Concrete Mixtures and Specifications for Today Making the Best Use of Local Materials to Reduce Costs, Concrete International, American Concrete Institute, MI, Vol. 24 No. 2, February 2002.
5. Crouch, L., Sauter, H., and Williams, J., 92-MPa Air-entrained HPC, TRB-Record 1698, Concrete 2000, p. 24, 2000.
6. Kailasananthan, K., McCullough, B.F., and Fowler, D.W., A Study of the Effects of Interface Condition on Thin Bonded PCC Overlays, Research Report 357-1, Center for Transportation Research, The University of Texas at Austin, October 1984.
7. Kosmatka, S., and Panarese, W., *Design and Control of Concrete Mixtures*, 13th ed., Portland Cement Association, 1992.
8. Menzel, C., Causes and Prevention of Crack Development in Plastic Concrete, in *Proceedings Annual Meeting of Portland Cement Association*, 1954.
9. Whitney, D., Isis, P., McCullough, B.F., and Fowler, D.W., Research Report 920-5, An Investigation of Various Factors Affecting Bond in Bonded Concrete Overlays, Center for Transportation Research, The University of Texas at Austin, Austin, TX, June 1992.
10. Suh, Y.C., Lundy, J.R., McCullough, B.F., and Fowler, D.W., A Summary of Studies of Bonded Concrete Overlays, Research Report 457-5F, Center for Transportation Research, The University of Texas at Austin, 1988.
11. Teo, K., Fowler, D.W., and McCullough, B.F., Monitoring and Testing of the Bonded Concrete Overlay on Interstate Highway 610 North in Houston, Texas, Research Report 920-3, Center for Transportation Research, The University of Texas at Austin, 1989.
12. Lundy, J.R., McCullough, B.F., and Fowler, D.W., Delamination of Bonded Concrete Overlays at Early Ages, Research Report 1205-2, Center for Transportation Research, The University of Texas at Austin, 1991.

Chapter 3
Condition Evaluation of the Existing Structure Prior to Overlay

L. Courard, M. Treviño and B. Bissonnette

Abstract The knowledge and analysis of the causes of degradations is the first step in repairing existing concrete structures. The assessment is performed by using destructive and non-destructive methods. These allow detecting modification of physical, chemical and/or mechanical properties of concrete: cracking, deformation, spalling, carbonation, etc., are various causes of distress, with potential irreversible consequences on concrete structure behaviour.

A description of specific methods for pavements and structural slabs is proposed; it is usually recommended to combine complementary methods for efficient diagnosis. This chapter wants to offer an overview of techniques and references that could help the user in selecting the most appropriate investigation program.

3.1 Introduction

The proper planning of investigations before carrying out repair works is important if optimum use is to be made of the test data. A major step needs to be taken from knowing, for example, that chloride is present in the concrete, to select the best methods for repair. Engineers require guidance on both the techniques available for the condition assessment of structures and the methods for data interpretation.

Two major stages are usually recommended for such an operation [1]. The first stage is based on a rapid-scan visual assessment, often including limited sampling in areas obviously damaged, from which areas can be selected for more detailed

L. Courard
GeMMe – Building Materials, ArGEnCo Department, University of Liège, Belgium

M. Treviño
Center for Transportation Research, The University of Texas at Austin (TX), U.S.A.

B. Bissonnette
Centre de Recherche sur les Infrastructures en Béton (CRIB), Université Laval, Québec (QC), Canada

investigations. The second stage contains a detailed diagnostic survey that relies on destructive and non-destructive testing techniques.

Non-destructive techniques are more and more utilized not only for the evaluation of the concrete strength but also for the detection of cracks and delaminations. More sophisticated techniques have been developed recently, mainly for the study of the concrete at early ages [2]. But they are easily applicable for the analysis of old structures. Finally, the interest for techniques able to evaluate concrete permeability is rising, due to the importance of this factor for durability evaluation.

The sections hereafter will discuss the assessment of pavement and deck/slab concrete structures. Even if the material is the same, there are differences between these two types of structures: actions, loading, aggressions, etc., that can lead to differences in the investigations and the interpretation of the results of the assessment.

3.2 Assessment of Pavement/Substrate Base

3.2.1 Principles of Evaluation

The first stage in the evaluation [3] of an existing pavement is the identification phase. It involves the following steps:

- assessment of the type of road structure with its environmental and loading conditions;
- visual examination of the surface layer of the road;
- coring to check the thickness and the residual performance of the concrete; and
- assessment of the possible causes of distress.

When this identification phase is completed, providing an insight into the possible origins of the problems, a more quantitative and problem-specific evaluation can be performed [3]. This second stage is the quantification stage.

3.2.2 Condition Survey of Distress

Two types of road structures are usually concerned: discontinuous concrete slabs and continuously reinforced concrete pavement overlays. The evaluation of the road structure is based on the following activities:

- visual examination;
- investigations;
- analysis of observations; and
- diagnostic.

Each step is important in order to have an accurate estimation of the state of the concrete in the road structure.

Table 3.1 Defaults in relation with road concrete degradations (CRR 1991)

Group	Type	Description
Cracking	Transverse crack	Rupture line perpendicular to the axis of the road
	Longitudinal crack	Rupture line parallel to the axis of the road
	Corner crack	Rupture line between transverse joint and longitudinal edge of the overlay
	Inclined crack	Short rupture line inside the slab and with 45° inclination with road axis
	Alligator cracking	Mesh cracking
Deformation	Pumping (punch-out)	Water or sludge pumping in the joint when vehicles passage
	Stair steps (faulting)	Difference of level between the two edges of the joint or the crack
	Settlement	Settlement of the edges of the slab, after breaking up
Spalling	Pot-hole	Hole with rounded shape and affecting concrete layer continuity
	Scaling	Superficial disintegration of the concrete
Joints distress	Opening of longitudinal joint	Large and irreversible opening of the joint
	Degradation (spalling)	Explosion of the upper part of the overlay around the joint
	Loss of waterproofing properties	Cracking or debonding or pull-out of the sealing mass
	Creep of sealing mass	Overflowing of the sealing mass out of the joint

3.2.2.1 Visual Examination

The general situation of the road must be checked (situation, drainage of water, profile, environment, etc.) before the observation of the quality of the concrete. The main types of degradations have been described by the Belgian Road Research Centre and are summarized in Table 3.1 (CRR 1991). The type and the situation of the defaults must be recorded. It is clear that the visual condition surveys not only involve mapping of the cracks, but also includes a visual survey of the drainage system. Information is required on the level of the road with respect to that of its surroundings and to the presence or absence of large slab faulting or other large unevenness.

3.2.2.2 Structural Investigations

Structural investigations are needed to give the data necessary for the choice and the design of the overlay. Analysis of the structure can be done in different ways:

- Inspection trenches along the road structure: thickness and aspect of the concrete layer, composition of the different layers of the road structure, type of sub-soil, etc.

- Coring: coring can give the same information as obtained from inspection trenches. Cores taken at cracks can give essential information on the potential for reflective cracking (orientation, evolution, propagation, etc.). The cores will also permit laboratory testing investigations to be conducted (see Section 2.3.).
- Falling Weight Deflection measurements (FWD) for the back calculation of E and E_S moduli. These measurements should be taken on sound areas, and preferably between the wheel tracks [3].
- Deflection measurements at joints or cracks to assess the load transfer across the crack or joint. These measurements should be done together with crack width measurements. They will be useful to estimate whether or not voids or loss of support have developed due to, for example, erosion and pumping.
- Crack activity measurements to determine in more detail the degree of slab rocking and the load transfer characteristics across the crack.

A structural analysis every kilometre of the road is a minimum (CRR 1991) if the visual inspection concludes to the homogeneity of the concrete upper layer. If not, it is necessary to select zones (good and bad) where more investigations are to be carried out: it can lead to five or more locations every kilometre of road.

3.2.2.3 Interpretation

Data coming from visual and structural inspections are analysed in order to point out the causes and to choose the solutions. These data are recorded in a matrix that should include the identification of the road section and, for each type of defaults (Table 3.1), the percentage of the concrete base affected or the cumulated length. Structural investigations – on site and in the laboratory – are also analysed and introduced in the same type of chart in such a way that zones can be classified and referred, regarding the intensity of the degradation. Finally, a diagnostic is proposed and potential causes of defaults and distress are pointed out, leading to the choice of the intervention: localized repair, overlay, slab replacement, or complete reconstruction of the road structure.

The Belgian Road Research Centre (CRR 1991) has proposed a strategy of rehabilitation, regarding the existing situation of the concrete roadway structure (Table 3.2).

Accordingly, the various possible rehabilitation procedures are classified in eight different categories (Table 3.3).

The characterization of the material itself can be conducted in a similar way than the one described for structural slabs and decks.

Table 3.2 Diagnostic and remedial measures (CRR 1991)

Visual observations	Condition of the substrate	Drainage	Number of vehicles per day		
			< 4,000	4,000 to 18,000	> 18,000
Isolated cracks affecting less than 10 % of the length of the road section. Low stair steps (< 3 mm). No settlement.	A or B	adapted	1	1	1
	A	unadapted	1	1 & 2	1 & 2
	B	unadapted	1 or 3		
Cracks affecting between 10 and 50 % of the length of the road section. Stair steps (< 6 mm). Low settlements.	A or B	adapted	1 or 3	1 or 3 or 4	5
	A	unadapted	1 or 3	1 & 2 or 2 & 3 or 2 & 4	2 & 5
	B	unadapted	1, 3 or 4	2 & 4 or 2 & 5	2 & 8
Cracks affecting more than 50 % of the length of the road section. Stair steps > 6 mm or important settlements.	A or B	adapted	6 or 7	6, 7 or 8	6, 7 or 8
	A	unadapted	6 or 7	2 & 6 or 2 & 7	2 & 6 or 2 & 7
	B	unadapted	2 & 6 or 2 & 8	2 & 6 or 2 & 8	2 & 8

Note: A = substrate is composed of quality and non polluted materials
B = substrate is composed of low quality and polluted materials

Table 3.3 Remedial measures (CRR 1991)

Code	Solution proposal
1	ontinuing usual maintenance
2	Increasing of drainage efficiency
3	Patch repairs
4	Replacement of deteriorated slabs
5	Thin adhesive overlay
6	Thin non-adhesive overlay
7	Replacement of the old overlay
8	Total reconstruction of the road structure

3.2.3 Test Methods for Pavement/Substrate Base

3.2.3.1 Deflection Testing

Since 1960, the non-destructive deflection testing methods have become more wide-spread known and have earned acceptance as a means to evaluate the in-situ pavement conditions [4].

The most common device for deflection evaluation is the Falling Weight Deflectometer (FWD), shown in Figure 3.1. There are other deflection devices that were used in the past, such as the Benkelman Beam, Dynaflect and Road Rater.

With the FWD, the loads are applied in several locations, by dropping a fixed weight from a given vertical height above the pavement; different weights are

Fig. 3.1 Falling weight deflectometer (FWD)

dropped at each location. The pavement response is measured as deflections by sensors (geophones) placed at fixed distances from the load along the pavement.

The deflection testing intervals vary depending on the project characteristics. It is common practice to conduct FWD measurements approximately every 120 m, but this number is to be taken only as a general guideline. The length of the project, its overall condition, and the availability of resources will ultimately determine the frequency and extensiveness of the deflection testing. Two types of deflection information should be collected: measurements should be taken along continuous slabs of pavement with no cracks between the deflection sensors, i.e., at the mid-span, for elastic layer moduli evaluation; the second type of measurement should be conducted across transverse cracks, for load transfer evaluation purposes. For this kind of measurement, it is recommended to arrange the FWD sensors with respect to the crack as illustrated in Figure 3.2, in which sensor number 4 is positioned on the other side of the crack with respect to the remaining sensors (downside arrangement).

3.2.3.2 Modulus of Elasticity

The modulus of elasticity of a pavement is a measure of its stiffness; it is the most common way of characterizing its load-carrying capacity. It is defined as the ratio of stress to strain. For materials with a linear stress-strain relationship, the modulus is simply the slope of the stress-strain line, but for non-linear materials like concrete, the modulus estimation is more complicated.

The non-linear behavior of concrete makes it obvious that the conventional concept of modulus of elasticity does not apply to concrete. Therefore, in order to systematically characterize this property, it is necessary to resort to arbitrary

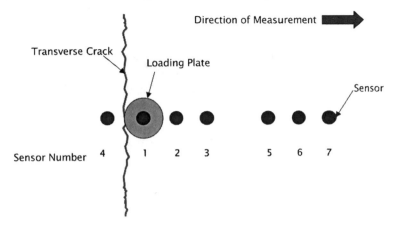

Fig. 3.2 FWD downside loading and sensor arrangement for load transfer measurement

definitions, based on empirical considerations. For instance, it is possible to define the initial tangent modulus or the modulus tangent to a predetermined point of the stress-strain curve. The secant modulus is more frequently used in laboratory tests.

However, the most frequently used method to arrive at moduli of elasticity values for the pavement structure is from deflection testing. The surface deflection data from FWD collected at the midspan of slabs, i.e., between cracks, are used to calculate the elastic moduli of the pavement layers, by a procedure called backcalculation. Normally, only one predetermined loading level of the FWD is considered, which means that only one of the weight drops is utilized. A load magnitude of approximately 9,000 pounds is commonly utilized, since it simulates the standard wheel load of an ESAL at one spot (AASHTO 1993).

Back-calculation is an iterative process that may be tedious and time-consuming, therefore, it is recommended to analyze the FWD data by means of computer programs. The goal of backcalculation is to estimate the pavement material layer stiffnesses, trying to find a set of parameters that correspond to the best fit of the measured deflection basins and minimizing the differences between the measured and the calculated deflection bowls. A computer program like MODULUS [5, 6] may be used for this kind of analysis. The program is intended for flexible pavements, but it can be successfully applied to PCC pavements as well. This program is designed to process data collected with FWD using a linear elastic procedure to generate a database of computed deflection bowls prior to the backcalculation process. The program iterates until the measured and computed deflections converge.

An alternative method is to backcalculate the elastic moduli using the charts and equation (AASHTO 1993).

Modulus values of concrete pavement may also be determined in the laboratory by sample testing. Underlying strata can be tested too, but this is seldom done, as deflection testing constitutes a more economical option. Many times, laboratory tests are used as a resource to confirm the certainty of the backcalculated moduli.

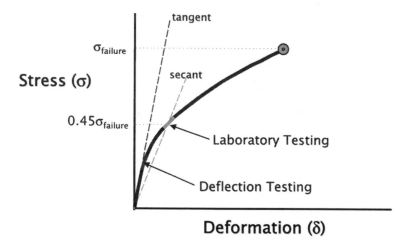

Fig. 3.3 Moduli of elasticity for coring and deflection testing

When both deflection and core testing are available and the resulting elastic moduli are compared, it should be noted that the moduli obtained from the core testing will be lower than the modulus from deflection tests because the stress level applied to the core samples is higher than the stresses that occur during deflection testing, and because of the non-linear elastic behavior of concrete. Typically, core specimens are loaded to approximately 45% of their estimated compressive strength and the secant modulus is calculated, whereas for deflection tests the stress levels are much lower, resembling an initial tangent modulus, as illustrated in Figure 3.3.

There are several empirical equations that have been developed to estimate the modulus from other concrete properties, like the expression shown below, where the concrete modulus, E_c, in psi, is inferred from the compressive strength, f'_c, in psi (1):

$$E_c = 57,000\sqrt{f'_c} \tag{3.1}$$

The same expression in SI units, i.e., with E_c and f'_c, in Pa, would be

$$E_c = 4,730,000\sqrt{f'_c} \tag{3.2}$$

3.2.3.3 Subgrade Modulus

From the backcalculation process, the moduli of all the pavement layers, slab, subbase and subgrade can be estimated. Nevertheless, the subgrade modulus requires additional considerations. Several factors affect the stress-strain relationship of soils, such as the moisture content, confining pressure and density, making it a very complex characteristic. When referring to subgrades, the elastic modulus of the soil is known as the resilient modulus, M_R. In rigid pavement design, besides the

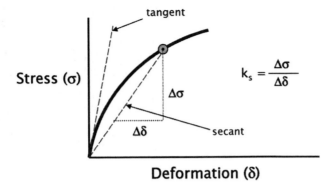

Fig. 3.4 Modulus of subgrade reaction

resilient modulus, another concept may be used to characterize the subgrade, known as the modulus of subgrade reaction, k.

The resilient modulus may be estimated by backcalculation of deflection test results, as explained above, which may be the easiest way, but it may also be determined in the laboratory using a triaxial test. The modulus of elasticity is the relationship between stress and strain; therefore, M_R is the ratio of the axial deviator stress, σ_d, and the recoverable axial strain, ε_a,

$$M_R = \frac{\sigma_d}{\varepsilon_a} \tag{3.3}$$

The modulus of subgrade reaction is defined as the ratio between an applied pressure and the ensuing deflection. It is determined by a plate-loading test. The plate is 30 in in diameter and the load is applied at a predetermined rate until a pressure of 10 psi is reached. The concept of the modulus of subgrade reaction is illustrated in Figure 3.4.

As Figure 3.4 shows, k varies depending on the stress level used, whether the modulus is considered tangent or secant. It also varies depending on the moisture conditions; therefore, it is recommended to analyze it on a seasonal basis. The modulus of subgrade reaction is directly proportional to the roadbed soil resilient modulus; however, there is no unique correlation between both. An empirical conversion from M_R, in psi, to k, in pci, (AASHTO 1993) is applicable only if the slab is placed directly on the subgrade, which does not occur very often:

$$k = \frac{M_R}{19.4} \tag{3.4}$$

The analogous expression in SI units, i.e., with M_R in Pa, and k in N/m^3 is

$$k = \frac{M_R}{0.5} \tag{3.5}$$

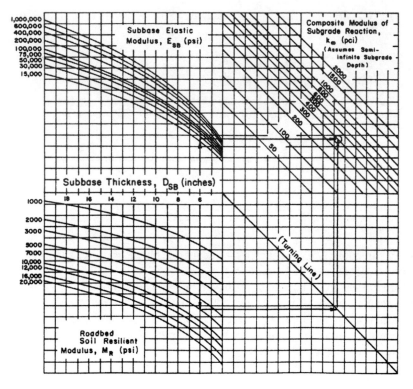

Fig. 3.5 AASHTO procedure to estimate the composite modulus of subgrade reaction

However, the previous equations render values of k that are too large, i.e., it overestimates k. A chart (AASHTO 1993) is proposed (Figure 3.5), for the cases in which the pavement slab is placed on top of a subbase, to determine the composite modulus of subgrade reaction, developed applying the 30-in plate loading onto a two-layer system.

3.2.3.4 Concrete Flexural Strength

Only for certain applications of concrete, one of which is PCC pavements, it is necessary to know its resistance under flexion. Also known as modulus of rupture, the flexural strength of PCC is evaluated by beam-breaking tests. Concrete beams, $6 \times 6 \times 18$ in, are cast and tests are normally performed at 7, 28 and 90 days. The beam specimens are tested in third point loading, with the break occurring suddenly with the appearance of a single crack. The value used for design is the mean value of the 28-day test specimens. An alternative way of estimating the modulus of rupture (6) is by means of an approximate relationship with the concrete compressive strength [7]:

$$S'_c = K\sqrt{f'_c} \qquad (3.6)$$

where S'_c is the PCC modulus of rupture; K is the constant between 8 and 10; and f'_c is the PCC compressive strength.

3.2.3.5 Load Transfer Efficiency (LTE)

The LTE of a pavement structure refers to its ability to transfer loads across transverse discontinuities such as joints or cracks. A high LTE value indicates that the pavement structure is capable of distributing the loads adequately at the discontinuities. The LTE calculations make use of the deflection data collected across transverse cracks as explained before, and as shown in Figure 3.2. In this study, two different procedures are recommended to calculate LTE, which make use of a deflection ratio. The first procedure is described by the following expression proposed by Teller [8]:

$$\text{LTE} = \frac{2W_U}{W_U + W_L} * 100\% \qquad (3.7)$$

where LTE is the load transfer efficiency (percentage), W_U the deflection on the unloaded slab and W_L the deflection on the adjacent loaded slab. If the LTE is zero, it means that no load is transferred from the loaded slab to the adjacent unloaded slab. In the case of perfect load transfer, the load is distributed completely from the loaded slab to the unloaded adjacent slab (i.e., the deflection is the same in both slabs) and LTE is equal to 100%. The second procedure for determining LTE was developed by Darter [8]:

$$\text{LTE} = \frac{W_U}{W_L} * 100\% \qquad (3.8)$$

where the variables are the same as in Teller's equation.

3.3 Assessment of Structural Slab/Deck/Support

3.3.1 Introduction

All assessments should be made of the defects in the concrete structure, their causes, and of the ability of the concrete structure to perform its function. The process of assessment should include, but not be limited to the following (ENV 1504-9, 4.3):

- present condition of the existing concrete structure, including non-visible and potential defects;
- original design approach;
- environment, including exposure to contamination;
- conditions during construction (including climatic conditions);
- history of the concrete structure;

- conditions of use (e.g. loading); and
- requirements for the future use of concrete structure.

he results of the completed assessment are only valid at the time that the repair works are designed and carried out. The nature and causes of defects, including potential combinations of causes, should be identified and recorded.

3.3.2 Condition Survey of Distress

In most cases assessment will have been carried out as a separate operation before the start of the protection or repair, but in all cases it is essential to assess the full extent and the causes of defects. Assessment may be undertaken in several stages. The purpose of a preliminary stage is to advise on the immediate safety of the concrete structure, to give an informed opinion on the urgency of commissioning further surveys, protection or repairs, and to outline what testing should be done to establish the causes and likely extent of the defects. The purpose of further assessment is as follows:

- to identify cause(s) of defects;
- to establish the extent of defects;
- to establish whether defects can be expected to spread to parts that are at present unaffected;
- to assess the effect of defects on structural capacity; and
- to identify all locations where protection or repair may be needed.

Assessment should include testing or other investigation with the aim of revealing hidden defects and causes of potential defects. The causes and type of defect should be considered during assessment. Causes may include, but are not limited to those listed below (ENV 1504-9) and in Table 3.4:

1. Causes of defects due to inadequate structural design.
2. Causes of defects due to inadequate construction or materials:

    - inadequate mix design, insufficient compaction, insufficient mixing, excess water in mix;
    - insufficient cover;
    - insufficient or defective waterproofing;
    - contamination, poor or reactive aggregates; and
    - inadequate curing and excessive evaporation.

3. Causes of defects revealed during service:

    - foundation movement, impacted movement joints, overloading; and
    - impact damage, expansion forces from fires.

4. External environment and agents:

Table 3.4 Common causes of defects (ENV 1504-9:1997)

Defects in concrete				Reinforcement corrosion	
Mechanical	Chemical	Physical		Corrosive contaminants	
				At mixing	From the environment
Impact Overload Movement (e.g. settlement) Vibration Explosion	Alkali-aggregate reaction Aggressive agents (e.g. sulphates, salts, fresh water) Biological activities	Freeze-thaw Thermal Salt crystallization Shrinkage Erosion Wear Fire	Carbonation and stray currents	Sodium chloride Calcium chloride	Sodium chloride Other contaminants Deicing salts Sea water

- severe climate, atmospheric pollution, chloride, carbon dioxide, aggressive chemicals;
- erosion, aggressive groundwater, seismic action; and
- stray electric currents.

The most common symptoms of concrete floor damage [9] are full-depth cracks or, less frequently, surface cracks. Another type of damage is delamination, displayed in the first stage by a hollow sound when tapping and breaking off floor panels in the final stage. The damages caused by chemical corrosion or freeze corrosion can be observed in places subject to aggressive environment influence. The freeze corrosion frequently occurs in parking decks. Floor defects include unevenness and lack of slope for drainage, which result in standing water. Concrete floors defects also include wash boarding and setting-off the joints as well as excessive abrasion and dusting.

For the repair methods involving the application of mortar and concrete, the European Standard EN 1504-10 recommends (Table 3.5) a series of investigations before and/or after preparation of the concrete substrate.

Table 3.5 gives a "memory-help chart", developed in order to avoid forgetting the main characteristics and properties of concrete to be involved in diagnosis. This information is fundamental and will be partially explained and developed into the next sections.

3.3.3 Structural Evaluation

It is sometimes necessary to evaluate the residual loading capacity of structural elements in order to choose the repair technique. Different types of investigations can be realized, in regards with the local situation of the structure:

Table 3.5 Summary of tests and observations for quality control (EN 1504-10:2002)

Test No.	Characteristic	Test Method or Observation	Standard(s)	Frequency	Status
1	Delamination	Hammer sounding (T)		Once before application	A
2	Cleanliness	Visual (O) Wipe test (T)		After preparation and immediately before application	A
3	Roughness	Visual sand test (O) or Profile Meter (T)	ISO 3274		B
4	Surface tensile strength of substrate	Pull-off test (T)	EN 1542 BS 1881 BS 207		B
5	Crack movement	Mechanical or electrical gages (O)	BS 1881 BS 206		C
6	Vibration	Accelerometer (O)			C
7	Temperature of the substrate	Thermometer (O)		Throughout application	A
8	Carbonation	Phenolphtalein test (T)	EN 104865		C
9	Chloride content	Site sampling and chemical analysis (T)	BS1181-124 DB5423.78 NTBUILD 208		C
10	Penetration of other contaminants	Site sampling and chemical analysis (T)			C
11	Electrical resistivity	Wenner test (T)			C
12	Compressive strength	Core and crushing test (T) Rebound hammer test (T)	EN 12504-1 EN 12504-2		B

T: test
O: observation
A: for all intended uses
B: for certain intended uses where required by the specific or operating conditions
C: for special applications

- tests on slab from the site if the structure is made of many repetitive elements;
- loading tests on site, in order to determine residual flexure or axial rigidity;
- dynamic evaluation tests, in order to evaluate rigidity from proper frequency measurement; and
- stress relaxation evaluation, with flat jacks and pressiometers into concrete sawed cracks.

It is also possible to evaluate the state of stress inside the concrete structure by means of flat jacks, double sawing method and hole core stress relaxation [10]. All these techniques have to be performed in order to attest the global stability of the structure before repair.

Generally speaking, the evolution of the geometry of the structural elements – also hidden – needs to be surveyed, and displacements, rotations, and crack growth

Table 3.6 Methods for the evaluation of concrete material

In-place tests to estimate strength	Non-destructive tests for integrity
Rebound hammer	Visual inspection
Ultrasonic pulse velocity	Stress wave propagation methods
Probe penetration	Ground penetrating radar
Pull-out	Electrical/magnetic methods
Break-off	Nuclear methods
Maturity method	Infrared thermography

have to be measured. For the global movements, geodetic methods, GPS as well as photogrammetry can be useful. In the case of the cracks, extensometers and gauges can follow their potential activity and give fundamental information on the cause(s) of defects.

3.3.4 Material Evaluation

Material evaluation can be deduced from destructive and non-destructive testing. There is, in fact, no standard definition for *non-destructive testing*. To some, they are tests that do not alter the concrete. For some, they are tests that do not impair the function of a structure, in which case the drilling of cores is considered as a NDT test.

An interesting classification of NDT has been proposed by Carino [11]. He proposes to divide the various methods into two groups (Table 3.6): (1) those whose main purpose is to *estimate strength*; and (2) those whose main purpose is evaluate conditions other than strength, i.e. to *evaluate integrity*.

Visual inspection is one of the most versatile and powerful of the NDT methods, and it is typically one of the first steps in the evaluation of a concrete structure. Visual inspection can provide a wealth of information that may lead to positive identification of the cause of observed distress. However, its effectiveness depends on the knowledge and experience of the investigator. Broad knowledge in structural engineering, concrete materials, and construction methods is needed to extract the most information from visual inspection. For example, cracking or spalling of a reinforced concrete structure or member may occur for many reasons (Table 3.4). But a simple visual inspection is insufficient to find out detailed information, such as whether reinforcement is corroding. Cracking of the concrete cover zone may be present without any obvious sign of further damage [1], such as rust staining or spalling. Equally, rust staining need not be accompanied by cracking of the cover, and occasionally may be due to other factors than corrosion of the reinforcing bar.

Useful guides are available to assist less experienced individuals in recognizing different types of damage and determining the probable cause of the distress (ACI 201.1R, ACI 207.3R, ACI 224.1R, ACI 362R).

A typical visual investigation might involve the following activities [11]:

- perform a walk-through visual inspection to become familiar with the structure;
- gather background documents and information on the design, construction, maintenance, and operation of the structure;
- plan the complete investigation;
- perform a detailed visual inspection; and
- perform any necessary sampling or in-place tests.

3.3.5 Test Methods and Procedures

3.3.5.1 Non-Destructive Testing for the Estimation of Strength

Schmidt rebound hammer

Due to its simplicity and low cost, the Schmidt rebound hammer is the most widely used non-destructive test device for concrete (related standard test method: *ASTM C805 Standard Test Method for Rebound Number of Hardened Concrete*). While the test appears simple, there is no simple relationship between the rebound number and the strength of concrete. In principle, the rebound is affected by the movement of the end of the plunger in contact with the concrete. The more the end of the plunger moves, the lower the rebound is. Thus the rebound number is likely to be influenced by the elastic stiffness and the strength of the concrete [11]. Since the rebound number is indicative of the near-surface properties of the concrete, it may not be indicative of the bulk concrete in a structural member. The results will be influenced by the moisture conditions, the carbonation, the surface texture and the orientation of the instrument [11, 12].

In summary, the rebound number method is recognized as a useful tool for performing quick surveys to assess the uniformity of concrete. However, because of the many factors besides concrete strength that can affect rebound number, it is not generally recommended where accurate strength estimates are needed.

Ultrasonic Pulse Velocity (UP-V)

The Ultrasonic Pulse Velocity method (related standard test method: *ASTM C597 Standard Test Method for Pulse Velocity Through Concrete*) is a stress wave propagation method that involves measuring the travel time, over a known path length, of a pulse of ultrasonic compressional waves (Figure 3.6). A timing circuit is used to measure the time it takes for the pulse to travel from the transmitting to the receiving transducers. The speed of compressional waves in a solid is related to the elastic constants (modulus of elasticity and Poisson's ratio) and the density. By conducting tests at various points on a structure, locations with lower quality concrete can be identified by their lower pulse velocity.

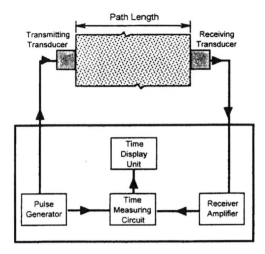

Fig. 3.6 Schematic of ultrasonic pulse velocity method

Moisture content, presence of reinforcement parallel to the pulse propagation direction, and presence of cracks or voids can affect the measured Ultrasonic Pulse Velocity. It should be used for estimating concrete strength only by experienced individuals. Like the rebound hammer test, the pulse velocity method is very useful for assessing the uniformity of concrete in a structure. It is often used to locate those portions of a structure where other tests should be performed or where cores should be drilled.

Probe Oenetration

The probe penetration method (Figure 3.7) involves using a gun to drive a hardened steel rod, or probe, into the concrete and measuring the exposed length of the probe (related standard test method: *ASTM C803 Standard Test Method for Penetration Resistance of Hardened Concrete*). In principle, as the strength of the concrete increases, the exposed probe length also increases; that means that the exposed length can be used to estimate compressive strength.

Probe penetration is not strongly affected by near-surface conditions, and is therefore not as sensitive as the rebound number method to surface conditions [11].

Pull-out Test

The pull-out test that is of interest here is the post-installed version. In this method, an insert with an enlarged head is sealed in the concrete. The insert and the accompanying conical fragment of concrete are extracted by using a tension loading device reacting against a bearing ring that is concentric with the insert. The force required

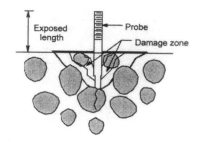

Fig. 3.7 Schematic of conical failure zone during probe penetration [11]

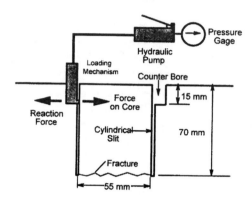

Fig. 3.8 Schematic of break-off test [11]

to pull out the insert is an indicator of concrete strength. Several techniques have been investigated in order to realize the anchorage of the insert [11].

Break-off Test

This test (related standard test method: *ASTM C1150 Standard Test Method for the Break-Off Number of Concrete*) measures the force required to break off a cylindrical core from the concrete mass (Figure 3.8).

The test subjects the concrete to a slowly applied force and measures a static strength of concrete. The correlations between break-off strength and compressive strength have been found to be non-linear [13], which is in accordance with the usual practice of relating the modulus of rupture of concrete to the square root of compressive strength.

Pull-off Test

The pull-off test is commonly used to estimate adhesion of repair systems on concrete structure (Figure 3.9).

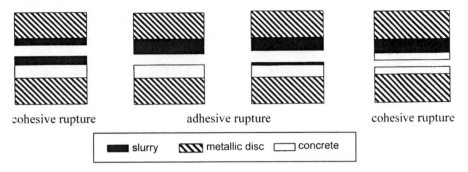

cohesive rupture   adhesive rupture   cohesive rupture

■ slurry   ⧄ metallic disc   ▭ concrete

Fig. 3.9 Effects of coring depth and diameter on the value of superficial cohesion of concrete [14]

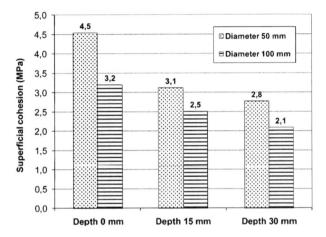

Fig. 3.10 Effects of coring depth and diameter on the value of superficial cohesion of concrete [15]

Also, it can be used to investigate the integrity and cohesion of a concrete surface to be overlaid. An experimental program has recently been conducted [15] to evaluate the influence of various test parameters on the measured cohesion of a reference concrete surface: transfer plate thickness and diameter, core drilling depth, speed of loading, adhesive type and thickness, and number of tests. A multivariate statistical analysis of the test results has shown that the plate diameter and core depth are the most significant parameters, presumably with threshold values, and that there exists a synergetic effect between them (Figure 3.10). Also, a minimum number of five tests can be defined, based on the standard deviation and required level of confidence of the results.

The results of this test will give not only an evaluation of the cohesion of the superficial layer of concrete, but also an estimation of the concrete integrity. Presence of cracks parallel to the surface, induced or not by surface preparation [16], will be easily detected. The related European Standard procedure (*EN1504-3 Products and Systems for the Protection and Repair of Concrete Structures – Part 3: Structural and Non-Structural Repairs*) requires a surface cohesion value of at least 1.0 MPa and 2.0 MPa for non-structural and structural repairs respectively. In any case, it should be equal to the recorded substrate tensile strength.

Optic Fibres

If accessible cavities are present, video endoscopes can be used for investigations. A fibre optic probe consisting in flexible optical fibres, lens and illuminating system is inserted into drilled hole, cracks, voids, etc. [12]. Observation of invisible parts of the structures can be performed, with the possibility of multidirectional observation and record by means of a camera.

3.3.5.2 Non-Destructive Tests for Flaw Detection and Condition Assessment

The other types of NDT methods are those for flaw detection and condition assessment. In this context the term *flaw* can include voids, honeycombing, delaminations, cracks, lack of sub-base support, etc. Recent research and development efforts on these methods have far exceeded those for methods to estimate strength [11]. The research impetus has come primarily from the transportation industry, since much of the highway infrastructure is need of repair as a result of natural aging or damage resulting from corrosion of reinforcing steel or deterioration of concrete.

The techniques for flaw detection are based generally on the following simple principle: the presence of an internal anomaly interferes with the propagation of certain type of waves. By monitoring the response of the test object when it is subjected to these waves, the presence of the anomaly can be inferred. The interpretation of the results of these types of NDT methods usually requires an individual who possesses knowledge both in concrete technology and in the physics governing the wave propagation.

Stress Wave Propagation Methods

Tapping an object with a hammer or steel rod (sounding) is one of the oldest forms of non-destructive testing based on stress wave propagation (Figure 3.11). The disturbance induced by impact propagates through the solid as three different waves: a P-wave (normal stress), a S-wave (shear stress) and an R-wave (Rayleigh waves). The P-wave and S-wave propagate into the solid along spherical wavefronts. In addition, there is an R-wave that travels away from the disturbance along the surface. In an infinite isotropic, elastic solid, the P-wave speed is related to the Young's modulus of elasticity, E, Poisson's ratio v, and the density, ρ. The reflected wave R is directly proportional to the specific acoustic impedance of the material and, consequently, to the modification of the interface, particularly when it is air (crack, hole, etc.).

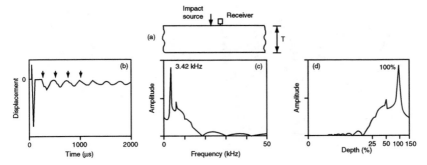

Fig. 3.11 Schematic figure of test using impact to generate stress waves with displacement wave-front, amplitude spectrum and normalized amplitude spectrum [17]

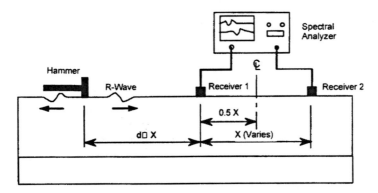

Fig. 3.12 Schematic figure of testing configuration for spectral analysis of surface waves

The method is subjective, as it depends on the experience of the operator, and it is limited to detecting near surface defects. Despite these inherent limitations, *sounding* is a useful method for detecting near-surface delaminations, and a related standard ASTM test procedure is available (ASTM D4580).

Figure 3.12 illustrates the disperse generalized Rayleigh surface wave (R-wave) spectrum analytical process. The received signal is analysed to obtain the dependence of phase velocity on the frequency [12].

The principle of *ultrasonic pulse-echo* (UP-E) is based on the action of an electro-mechanical transducer that generates a short pulse of ultrasonic stress waves that propagates into the object being inspected [18]. Reflection of the stress pulse occurs at boundaries separating materials with different densities and elastic properties (these determine the acoustic impedance of a material) [19, 20]. Numerous parameters limit the applicability of the method: the structures cannot be too complex, the defects not too short or far from the surface. Moreover, the network will sometimes limit the surface of investigation. The results actually obtained seem more efficient for the detection of pop outs and voids, or for the global modification of concrete characteristics.

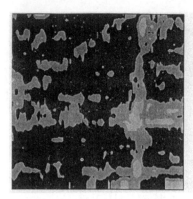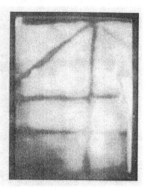

Fig. 3.13 Comparison between the results of the GPR and thermography tests (to locate position of large voids and inclusions of different materials) [21]

Infrared Thermography

Infrared thermography is a technique for locating near-surface defects by measuring surface temperature (related standard test method: *ASTM D4788 Standard Test Method for Detecting Delaminations in Bridge Decks Using Infrared Thermography*). The presence of an anomaly having a lower thermal conductivity than the surrounding material will interfere with the flow of heat and alter the surface temperature distribution. As a result, the surface temperature will not be uniform. Thus, by measuring the surface temperature, the presence of the defect can be inferred. In practice, the surface temperature is measured with an infrared scanner that works in a manner similar to a video camera. The output of the scanner is a *thermographic image* of temperature differences (Figure 3.13).

The procedure for performing an infrared thermographic survey to detect delaminations in bridge decks is described in ASTM D4788 [11]. Infrared thermography is a *global* inspection method which needs experts for interpretation of the results [11]. This permits large surface areas to be scanned in a short period of time, which is an advantage over other methods that have been discussed.

Ground Penetrating Radar (GPR)

Radar (acronym for **RA**dio **D**etection **A**nd **R**anging) is analogous to the ultrasonic pulse-echo technique previously discussed, except that pulses of electromagnetic waves (short radio waves or microwaves) are used instead of stress waves (Figure 3.14).

The pulse travels through the test object and when it encounters an interface between dissimilar materials, some of the energy is reflected back toward the antenna as an *echo*. By measuring the time from the start of the pulse until the reception

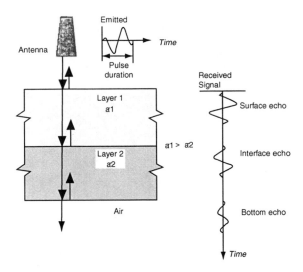

Fig. 3.14 Reflection of radar pulse at interfaces between materials with different relative dielectric constants and antenna signal caused by arrival of echoes [11]

of the echo, it is possible to determine the depth of the interface if the propagation speed through the material is known [11].

This method (*ASTM D4748 Standard Test Method for Determining the Thickness of Bound Pavement Layers using Short-Pulse Radar*) can lead to the detection of delamination and often for locating reinforcing bars in the structure [22]. Results seem to depend on moisture and chloride concentration in concrete.

Electrical and Magnetic Methods for Reinforcement

Information about the quantity, location, and condition of reinforcement is needed to evaluate the strength of reinforced concrete members. Magnetic and electrical methods are used to yield this information about embedded steel reinforcement.

Cover meters
Devices to locate reinforcing bars and estimate the diameter and depth of cover are known as *cover meters*, or *pachometers*. These devices are based upon interactions between the bars and low frequency electromagnetic fields. Commercial cover meters can be divided into two classes: those based on the principle of *magnetic reluctance* and those based on *eddy currents*. Cover meters are effective in locating individual bars.

By using multiple measurements methods, the bar diameter can generally be estimated to within two adjacent bar sizes if the spacing exceeds certain limits that are also dependent on the particular meter. But when multiple bars are present within the detection range of the instrument (Figure 3.15), the increased signal received

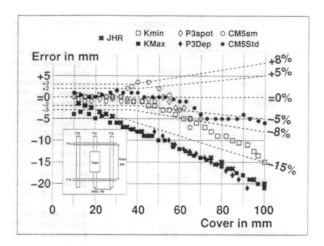

Fig. 3.15 Precision of cover meter measurements [23]

from the increased mass of steel invariably causes the instrument to indicate a cover which is shallower than the true cover [23]. This results in a lower accuracy of the results (Figure 3.15), depending on the type of the test device used.

Corrosion activity
Electrical methods are used to evaluate corrosion activity of steel reinforcement. The *half-cell potential method* includes a copper-copper sulfate half-cell, connecting wires, and a high-impedance voltmeter. The positive terminal of the voltmeter is attached to the reinforcement and the negative terminal is attached to the copper-copper sulfate half-cell. A high impedance voltmeter is used so that very little current flows through the circuit. The half-cell makes electrical contact with the concrete by means of a porous plug and a sponge that is moistened with a wetting solution (such as liquid detergent). There are numerous limitations to be considered when planning corrosion rate testing [24]:

- the concrete surface has to be smooth (not cracked, scarred, or uneven);
- the concrete surface has to be free of water, impermeable coatings or overlays;
- the cover depth has to be less than 100 mm;
- the reinforcing steel can not be epoxy coated or galvanized;
- the steel to be monitored has to be in direct contact with the concrete;
- the reinforcement has not to be cathodically protected;
- the reinforced concrete has not to be near areas of stray electric currents or strong magnetic fields;
- the ambient temperature has to be between 5 and 40°C;
- the concrete surface at the test location must be free of visible moisture; and
- test locations must not be closer than 300 mm to discontinuities, such as edges and joints.

Nuclear (Radioactive) Methods

Nuclear (or radioactive) methods for non-destructive evaluation of concrete involve the use of high-energy electromagnetic radiation to gain information about the internal structure of the test object (related standard test methods: *ASTM C1040 Standard Test Methods for Density of Unhardened and Hardened Concrete in Place by Nuclear Methods; BS 1881: Part 205 Recommendations for radiography of Concrete*). These involve a source of penetration electromagnetic radiation and a sensor to measure the intensity of the radiation after it has travelled through the object. If the sensor is in the form of special photographic film, the technique is called *radiography*. If the sensor is an electronic device that converts the incident radiation into electrical pulses, the techniques is called *radiometry*.

Radiometric methods
There are two basic radiometric methods that use X-rays and gamma rays in non-destructive testing of concrete. In the *transmission* method, the amplitude of the radiation passing through a member is measured. As the radiation passes through a member, the attenuation is dependant on the density of the material. Direct transmission techniques can be used to detect reinforcement. However, the main use of the technique is to measure the in-place density, both in fresh and hardened concrete. *In the backscatter-method*, a radioactive source is used to supply gamma rays and a detector close to the source is used to measure the backscattered rays. Backscatter techniques are particularly suitable for applications where a large number of in-situ measurements are required. Since backscatter measurements are affected by the top 40 to 100 mm, the method is best suited for measurement of the surface zone of a concrete element. A good example of the use of this method is the monitoring of the density of bridge deck overlays.

Radiography methods
Radiography provides a radiation-based photograph of the interior of the concrete. From this photograph, the location of the reinforcement, voids in concrete, or voids in grouting of post-tensioning ducts can be identified. The radiation passing through the test object is attenuated by different amounts depending on the density and the thickness of the material that is traversed. The main purpose of these methods is to *see* into concrete but they require extensive precautions, skilled personnel and highly sophisticated equipment.

3.3.6 Methods for the Determination of Superficial Porosity

3.3.6.1 Principles

Three major transfer mechanisms may occur in porous media, such as concrete substrate to be overlaid: diffusion, permeation and capillary suction [25]. Capillary suction, one of the main phenomena in creating contact, is directly in relation with the superficial porosity of concrete and is governed by Washburn's equation [26] which gives the change of height penetration of a liquid into a capillary as the square root of time and radius:

$$\ell_p = \sqrt{\frac{r \cdot \gamma_L \cdot \cos\theta}{2\eta} \cdot t} \tag{3.9}$$

where ℓ_p is the change of the height penetration of the liquid, γ_L is the surface free energy of the liquid, r is the radius of the pore, θ is the contact angle, η is the viscosity of the fluid and t is the time. If we assume a porous material to have a large number of capillaries parallel to each other, with different diameters and perpendicular to the surface of the liquid, the liquid will be absorbed with a higher force by the shorter capillaries, according to Laplace [27]), but faster and in large quantities by the largest pores, according to Washburn. It has been shown that the capillary absorption can be increased by selecting a liquid characterized by a surface free energy γ_L, just below the critical surface tension of wetting γ_c [25, 26].

The influence of the pore's entry is also clearly observed and will influence penetration rate [28]. Evaluation of the porosity of the substrate is consequently important for the appreciation of the situation. Capillary action test methods as well as more sophisticated analysis can be useful in this aim.

3.3.6.2 Capillary Suction Coefficient

The most commonly used test to analyse water transfer at the interface is the capillary suction test [28, 29]. The capillary suction test is described by several standards around the world: NBN B 14-201 (Mortars), DIN 52617, prEN 13057 (Repair products), EN 480-5 (Concrete admixtures) differ essentially by the water level above the bottom surface of concrete specimen and the time when measurement is taken. Mass change is usually measured after 5, 15, 30 and 45 minutes, as well as after 2, 6 and 24 hours. Mass is measured on samples wiped off with a damp tissue. From the capillary suction test it is possible to calculate the rate of water absorbed by the capillaries ($E_c\%$), in regards with the water absorbed under vacuum ($E_v\%$). One can then calculate the relative impregnation ratio S_t, at time t_i:

$$S_t = \frac{E_c}{E_v} \cdot 100 \tag{3.10}$$

This value is characteristic of the absorption rate of the substrate at any time vs. a complete filling of the porous skeleton (corresponding to accessible porosity of concrete samples). Classical and standardized tests are, however, not able to characterise the early behaviour of the composite concrete substrate/repair material because the time interval defined for observations begins too late and is too large. Adaptation of this test has been realized [28] to continuously record mass change of concrete samples as of early contact with the liquid phase and evolution over time of liquid capillary suction (Figure 3.16).

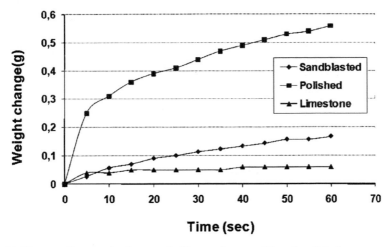

Fig. 3.16 Water capillary suction between 0–60 seconds on sandblasted, polished concrete surface and limestone [28]

An impregnation rate calculated between 0 and 20 seconds is an interesting discriminator parameter for the evaluation of transport kinetics and liquid migration at the interface. Figure 3.11 illustrates the different behaviour of samples prepared by sandblasting and polishing, as well the low absorption rate of limestone [28]. This test, however, requires that core samples be taken from the concrete, in order to perform the test in the laboratory. Knowledge of the structural skeleton of the voids can be completed by Mercury Intrusion Porosimetry.

3.3.6.3 Mercury Intrusion Porosimetry

Analysis of Mercury Intrusion Porosimetry (MIP) and nitrogen adsorption isotherms is based on pore shape, i.e. cylindrical, ink-bottle, etc. [30]. They concern low-radius capillaries, i.e. between 2 nm and 75 μm, unable to absorb large quantities of water in a short period of time [28]. Only superficial concrete (5 ± 1 mm) is analyzed, as it will be principally influenced by liquid movements. Mercury Intrusion Porosimetry (2000 bars) gives granulometry gradation of the pores between 7.5 nm and 75 μm (Figure 3.17), as well as evaluation of specific surface, total porous volume and mean radius.

Consequently, the potential effect of capillary entry shape on capillary action in repair techniques needs to be simultaneously analysed by means of MIP and classical absorption tests.

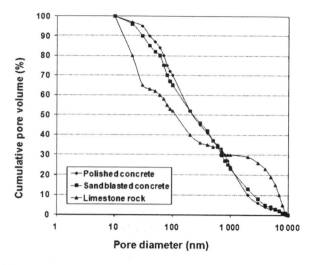

Fig. 3.17 Evolution of pore size distribution in the superficial layer of concrete surface prepared with sandblasting and polishing, in comparison with limestone aggregate [28].

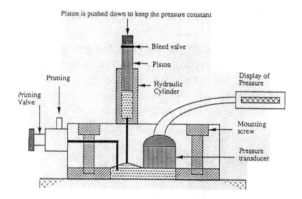

Fig. 3.18 In-situ permeability test – Method operation for water flow tests

3.3.6.4 In-situ Tests

If it is impossible to core samples from the structure, some tests exist and permit to qualify the porosity directly on the site. These systems can be used to measure the air and water permeabilities, and the water absorption (sorptivity). In this last case, the volume of water penetrating into the concrete is recorded at a constant pressure of 0.2 bars. As water introduced by a syringe is absorbed by capillarity, the pressure inside – under the piston – tends to decrease; hence it is maintained constant by pushing the piston through the cylinder (Figure 3.18). A plot of the quantity of water absorbed and the square root of time elapsed is linear and the slope of this graph represents the sorptivity index (m^3/min).

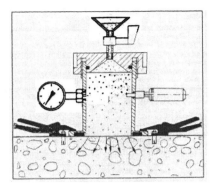

Fig. 3.19 Water permeation test

The same type of test device is based on a watertight gasket, glued on the concrete substrate (Figure 3.19). The valves of the housing are opened and the chamber is filled with boiled water. The valves are closed and the top lid of the housing turned until a desired pressure is achieved (between 0 and 6 bars). The pressure selected is maintained with a micrometer gauge pressing a piston into the chamber substituting the water penetrating in the concrete. This system can be used for evaluating the extent of microcracking and porosities of the concrete "skin" of the finished structure.

3.3.6.5 Methods for the Determination Chemical Damages

The main cause of reinforced concrete deterioration in service is corrosion of the reinforcement, due to a variety of causes (Figure 3.20). Depending on the processes at work, the following tests are routinely used in combination: depth of carbonation and cover, and chloride content and half-cell potential [1].

The depth of carbonation, by using phenolphthalein, is simply assessed either on site or in the laboratory, followed by cover depth. This will establish the relationship between depth of carbonation and cover around the structure for a given age, in order to define areas of risks.

Chloride ion content can be determined rapidly and inexpensively from the analysis of dust from drilled samples taken from concrete. However, chloride contents can be highly variable around a structure. Measurement of half-cell potential can be used to find suspect areas, from which dust drilling samples can be taken.

In special environments (sewage plants, agricultural, marine and food infrastructures, etc.), other chemical investigations are needed to assess the presence of contaminants in the concrete.

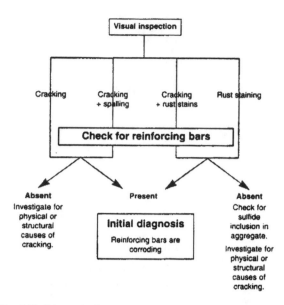

Fig. 3.20 Diagnosis flow chart for corrosion of reinforcement [1]

3.4 Feasibility Analysis

It is known that the values measured from NDT, which characterize only one particular physical property of the concrete, are influenced in different ways by the constituents of concrete and curing conditions. For this reason [31], it is important to combine different testing methods (Figure 3.21).

With regard to the properties that have to be determined, it is often interesting to first assess all the structures, in such a way that it is possible afterwards to point out zones where more investigations are needed. The NDT methods are particularly interesting in this aim: for example, UP-V and rebound hammer are usually combined. Results of tests on surface of the concrete member are combined with those concerning its whole thickness. The RILEM TC 43 CND [32] has provided recommendations about the essential criteria for the establishment of beneficial combinations (ibid., p. 43) and examples are given about potential combinations of two or three tests (ibid. pp. 47–48).

For sure, the strength relationship should be developed experimentally before using a test method to estimate in-place strength, by comparison with destructive tests on cores from different locations. Finally, the importance of having qualified operators cannot be overemphasized because NDT are indirect methods by which the property or characteristic of primary interest is inferred by measuring other properties or characteristics. A lack of understanding of the underlying principles and the interferences associated with these methods [11] could lead to incorrect assessment of concrete structures.

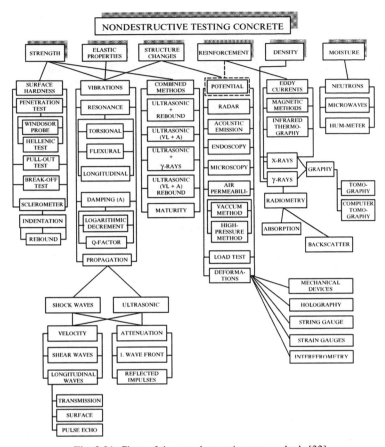

Fig. 3.21 Chart of the non-destructive test methods [32]

References

1. Robery, P.C., Investigation methods utilizing combined NDT techniques, *Construction Repair* **9**(6), 11–16, 1995.
2. Ballivy, G. and Rhazi, J., Progrès récents dans le domaine de l'auscultation. Auscultation et Instrumentation des Ouvrages de Génie Civil, Faculté des Sciences Appliquées de Béthune, 2000.
3. Molenaar, A.A.A. and Potter, J., Assessment of the reflection crack potential, in *Prevention of Reflective Cracking in Pavements*, Rilem Report 18, Chapter 3, E. and F.N. Spon, London, pp. 28–41, 1997.
4. Huang, Y.H., *Pavement Analysis and Design*, Prentice-Hall, Englewood Cliffs, NJ, 1993.
5. Uzan, J., Scullion, T., Michalak, C.H., Paredes, M., and Lytton, R.L., A microcomputer based procedure for backcalculating layer moduli from FWD data, Research Report 1123-1, Texas Transportation Institute, Texas A&M University, July 1988.
6. Scullion, T. and C. Michalak, C., Modulus 4.0: User's manual, Research Report 1123-4, Texas Transportation Institute, Texas A&M University, January 1991.
7. Yoder, E.J. and Witczak, M.W., *Principles of Pavement Design*, John Wiley & Sons, New York, NY, 1975.

8. Wei, C. and McCullough, B.F., Development of load transfer coefficients for use with the AASHTO guide for design of rigid pavements based on field measurements, Research Report 1169-3, Center for Transportation Research, The University of Texas at Austin, February 1992.

9. Czarnecki, L., Mechanisms of industrial floors deterioration – Causes, results and preventive means. In *Proceedings Fifth International Colloquium Industrial Floors '03*, Esslingen, Germany (Technische Akademie Esslingen), P. Seidler (Ed.), pp. 65–72, 2003.

10. Plumier, A., Pathologie et réparations structurelles des constructions – Partie 2: Caractérisation d'une structure existante, Département Mécanique des Matériaux et des Structures, Faculté des Sciences Appliquées, Université de Liège (notes de cours), 38 pp., 1994.

11. Carino, N.J., Non destructive test methods to evaluate concrete structures. In *Proceedings Sixth CANMET/ACI International Conference on the Durability of Concrete*, Special Seminar, Thessaloniki, Greece, p. 75, 2003.

12. Garbacz, A. and Garboczi, E., Ultrasonic evaluation methods applicable to polymer concrete composites, US Department of Commerce, Report No. NISTIR 6975, Gaithersburg, MD, 2003.

13. Barker, M.G. and Ramirez, J.A., Determination of concrete strengths with break-off tester, *ACI Materials Journal*, **85**(4), 221–228, 1988.

14. Courard L., Adhesion of repair systems to concrete: Influence of interfacial typology and transport phenomena, *Magazine of Concrete Research*, **57**, 1–10, 2005.

15. Courard, L. and Bissonnette, B., Essai dérivé de l'essai d'adhérence pour la caractérisation de la cohésion superficielle des supports en béton dans les travaux de réparation: Analyse des paramètres d'essai, *Materials and Structures* **37**, 342–350, 2004.

16. Bissonnette, B., Courard, L., Vayburd, A., and Belair, N., Concrete removal techniques: Influence on residual cracking and bond strength, *Concrete International*, December 2006.

17. Pratt, D. and Sansalone, M., Impact-echo signal interpretation using artificial intelligence, *ACI Materials Journal*, **89**(2), 178–187, 1992.

18. Cheng, C. and Sansalone, M., The impact-echo response of concrete plates containing delaminations: numerical, experimental and field studies, *Materials and Structures*, **26**(159), 274–285, 1993.

19. Garbacz, A., Czarnecki, L., and Clifton, J.R., Non-destructive methods to assess adhesion between polymer composites and Portland cement concrete. In *Proceedings of the Second International Rilem Symposium on Adhesion between Polymers and Concrete (ISAP'99)*, Dresden, Germany, M. Puterman and Y. Ohama (Eds.), Rilem Publications, pp. 467–474, 1999.

20. Dondonné, E. and Toussaint, P., Synthèse sur l'utilisation de l'impact-echo par la Direction de l'Expertise des Structures du Ministère Wallon de l'Equipement et des Transports, *Bulletin des Laboratoires des Ponts et Chaussées* (Special issue: Evaluations non destructives pour le génie civil), **239**, 51–62, 2002.

21. Binda, L., Saisi, A., and Tiraboschi, C., Investigation procedures for the diagnosis of historic masonries, *Construction and Building Materials*, **14**, 199–233, 2000.

22. Rhazi, J., Laurens, S., and Ballivy, G., Insights on the GPR non destructive testing method of bridge decks, Auscultation et Instrumentation des Ouvrages de Génie Civil, Faculté des Sciences Appliquées de Béthune, 2000.

23. Alldred, J.C., Quantifying losses in cover meter accuracy due to congestion of reinforcement, *Construction Repair*, **9**(1), 41–47, 1995.

24. Cady, P.D. and Gannon, E.J., *Condition Evaluation of Concrete Bridges to Reinforcement Corrosion. Volume 8: Procedure Manual*, SHRP-S/FR-92-110, National Research Council, Washington, DC, 124 pp., 1992.

25. Courard, L., Parametric study for the creation of the interface between concrete and repairs products, *Materials and Structures*, **33**(225), 65–72, 2000.

26. Kinloch, A.J., *Adhesion and Adhesives: Science and Technology*, Chapman and Hall, London, 441 pp., 1987.

27. Jouenne, C.A., Capillarité et tension superficielle. In *Traité de céramique et matériaux minéraux*, Septima, Paris, pp. 513–575, 1980.

28. Courard, L. and Degeimbre, R., A capillary action test for the investigation of adhesion in repair technology, *Canadian Journal of Civil Engineering*, **30**(6), 1101–1110, 2003.
29. Justnes, H., Capillary suction of water by polymer cement mortars. In *Proceedings of the RILEM Symposium on Properties and Test Methods for Concrete-Polymer Composites (ICPIC)*, Oostende, D. Van Gemert (Ed.), pp. 29–37, 1995.
30. Lecloux, A., Texture of catalysts. In *Catalysis: Science and Technology*, J.R. Anderson and M. Boudart (Eds.), Springer Verlag, Berlin, pp. 171–227, 1981.
31. Teodoru, G.Y.M. and Mommens, A., Non-destructive testing in the quality control of buildings: Why, what and how? In *Proceedings for the International Symposium on Quality Control of Concrete structures*, Gent, L. Taerwe and H. Lambotte (Eds.), E & F.N. Spon, pp. 367–376, 1981.
32. Facaoaru, R.C.L., Rilem draft recommendation for in-situ concrete strength determination by combined non-destructive methods, 43-CND Committee, *Materials and Structures*, **26**(155), 43–49, 1993.

Standards

AASHTO Guide for Design of Pavement Structures, American Association of State Highway and Transportation Officials, 1993.
American Concrete Institute, ACI Committee 201.1R, Guide for Making a Condition Survey of Concrete in Service.
American Concrete Institute, ACI Committee 207.3R, Practices for Evaluation of Concrete in Existing Massive Structures for Service Conditions.
American Concrete Institute, ACI Committee 224.1R. Causes, Evaluation and Repair of Cracks in Concrete Structures.
American Concrete Institute, ACI Committee 228.1R, In-Place Methods for Determination of Strength of Concrete, 1989.
American Concrete Institute, ACI Committee 228.1R, In-Place Methods to Estimate Concrete Strength, 1995.
American Concrete Institute, ACI Committee 228.2R, Non Destructive Test Methods for Evaluation of Concrete in Structures.
American Concrete Institute, ACI Committee 362R, Sate-of-the-Art Report on Parking Structures.
American Concrete Institute, ACI Committee 437 R, Strength Evaluation of Existing Concrete Buildings.
American Concrete Institute, Surface Roughness.
American Concrete Institute, ACI Committee 546. ACI Manual of Concrete Practice – Part 4: Concrete Repair Guide, 41 pp., 1998.
Belgian Standard NBN B 14-201, Test on mortars – Capillary suction test. Institut Belge de Normalisation, Brussels, 1973.
Belgian Standard NBN B 15-225, Essais des bétons – Mesure de la dureté de surface – Indice sclérométrique. Institut Belge de Normalisation, Brussels, 1984.
Belgian Standard NBN B 15-229, Essais des bétons – Essais non destructifs – Mesure de la vitesse du son. Institut Belge de Normalisation, Brussels, 1991.
Belgian Guidelines, G0007, Agreement and Certification Guidelines for Hydraulic Binder-Based Repair Mortars. Belgian Union for Technical Agreement in Construction (UBAtc), Sector Civil Engineering, Walloon Ministry of Equipment and Transportation, Liège, Belgium, 2000.
British Standard BS 1881: Part 207, Recommendations for the Assessment of Concrete Strength Near-to-Surface Tests. British Standards Institution, London, 1992.
Code de bonne pratique pour le renforcement des chaussées à l'aide du béton de ciment, Recommandations CRR – R63/91, Brussels, 1991.

Code de bonne pratique pour le dimensionnement des chaussées à revêtement en béton de ciment, Recommandations CRR – R57/85, Brussels, 1985.

German Standard DIN 52617, Determination of the γ coefficient of Building materials. Bestimmung des Wasseraufnamekoeffizienten von Baustoffen, Deutsches Institut für Normung, Berlin, 1987.

European Standard ENV1504-9, Products and systems for the protection and repair of concrete structures – Definitions, requirements, quality control and evaluation of conformity – Part 9: General principles for the use of products and systems. Comité Européen de Normalisation, Brussels, 1997.

European Standard prEN1504-10, Products and systems for the protection and repair of concrete structures – Definitions, requirements, quality control and evaluation of conformity – Part 10: Site application of products and systems and quality control of the works. Comité Européen de Normalisation, Brussels, 2002.

European Standard prEN 1766, Produits et systèmes pour la protection et la réparation des structures en béton – Méthodes d'essais – Bétons de référence pour essais, Comité Européen de Normalisation, Brussels, 1995.

European Standard prEN 13036-1, Surface characteristics. Test methods – Part 1: Measurement of pavement surface macrotexture depth using a volumetric patch technique. European Committee for Standardisation, Brussels, 2000.

European Standard prEN 13057, Products and systems for the protection and the repair of concrete structures – Test methods – Measurement of capillary absorption. European Committee for Standardisation, Brussels, 2000.

European Standard EN 480-5, Admixtures for concrete, mortar and grout – Test methods – Part 5: Determination of capillary absorption. European Committee for Standardisation, Brussels, 1996.

Chapter 4
Bond

J. Silfwerbrand, H. Beushausen and L. Courard

Abstract Good bond is a key factor for providing monolithic action in bonded concrete overlays. This chapter starts with three theoretical sections classifying bond in three groups (complete, uncertain, and poor), defining bond strength, and describing the fundamental bond mechanisms. The main part of this chapter covers a thorough description in chronological order of how the 13 most important factors affect bond. All events from removal of deteriorated concrete to concrete placing and curing and long-term exposure are investigated. There exist a number of different methods to determine bond strength and the most frequent ones are briefly described in a separate section. It is shown that there is an evident relationship between the two most frequently used families of test methods, i.e., between methods determining bond strength in tension and bond strength in shear. Furthermore, it is shown that it is possible to provide durable bond in concrete overlays if all operations for concrete removal, surface cleaning, concrete placing, and curing are conducted meticulously. The chapter is ended by two sections devoted to design strength values in various international codes and performance requirements that can be used for quality control.

4.1 Classification of Bond

Bond may be classified into three groups: (i) complete bond, (ii) uncertain bond, and (iii) poor bond or debonding (Figure 4.1). Complete bond is a prerequisite for

J. Silfwerbrand
Swedish Cement and Concrete Research Institute (CBI), SE-100 44 Stockholm, Sweden

H. Beushausen
University of Cape Town, Department of Civil Engineering, Cape Town, South Africa

L. Courard
GeMMe – Building Materials, ArGEnCo Department, University of Liège, Belgium

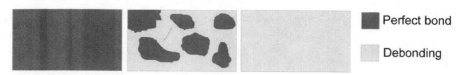

Fig. 4.1 Classification of bond into three groups: complete bond (left), uncertain bond (centre), and debonding (right)

structural interaction and the possibilities to consider at two-layer system as a monolithic one. Uncertain bond is the most difficult type. Since the overlay is bonded to the substrate in certain areas, a high restraint is developed. This restraint may together with overlay shrinkage lead to the development of wide cracks perpendicular to and between two areas of bonding. In cases with 100% debonding, the overlay may be considered as a slab on (stiff) grade.

4.2 Definition of Bond Strength

The aim of a concrete repair is to restore the load carrying capacity and the stiffness of a concrete structure or member. Consequently, monolithic action is the final objective. A prerequisite for monolithic action is sufficient bond between the remaining concrete and the new-cast overlay. The characteristics of adhesion, or bond, can be perceived from two different angles [1]: firstly, the conditions and kinetics of joining two materials, taking into account different bond mechanisms; and secondly, the quantitative measure of the magnitude of adhesion, usually expressed in stress or energy required to separate the two materials. For concrete overlays, bond strength is usually defined as the tensile strength perpendicular to the interface plane. Bond strength in shear may also be considered, given its inherent role and intrinsic relationship with the tensile behaviour [2], but tensile bond strength is much easier to measures (see Section 4.5).

In practice the bond strength is determined by coring and pull-off tests. A failure stress can be computed by dividing the maximum force by the area of the cross section. It is important to note that the bond strength solely equals this failure stress if the failure occurs completely at the interface. In all other cases (failures completely or partly in substrate or overlay), the failure stress is only a lower bound of the bond strength.

4.3 Fundamental Bond Mechanisms

First, adhesion has to be clearly defined because of the "duality" of the term [3]:

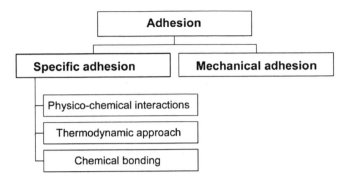

Fig. 4.2 Principles of the theory of adhesion

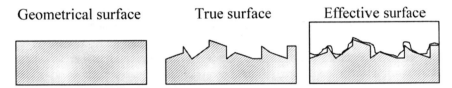

Fig. 4.3 Geometrical, true, and effective surface areas between substrate and overlay

on one hand, adhesion is understood as a process through which two bodies are brought together and attached – bonded – to each other, in such a way that external force or thermal motion is required to break the bond. On the other hand, we can examine the process of breaking a bond between bodies that are already in contact. In this case, as a quantitative measure of the intensity of adhesion, we can take the force or the energy necessary to separate the two bodies.

Adhesion has therefore two different aspects, according to whether our interest is mainly (1) in the conditions and the kinetics of contact or (2) in the separation process. The intensity of adhesion will depend not only on the energy that is used to create the contact, but also on the interaction existing in the interface zone [4]. The mechanism of adhesion can be classified in two phases, as presented in Figure 4.2.

Each individual model of adhesion can contribute to an explanation of the phenomenon, but is unable by itself to give a comprehensive explanation [5].

Bond mechanisms all depend on the true surface area, as opposed to the geometric surface area, and the contact surface area, also termed effective surface area (Figure 4.3). The effectiveness of mechanical adhesion [6] is explained by the fact that the liquid will penetrate through the roughness of the substrate and, after hardening, will induce cohesion by interlocking effect. Special preparation technique can increase the true surface area [7] and contribute to the increase of potential interaction sites. It ought to be mentioned that the true surface area is difficult to measure, not least since it is dependent on the fineness of the measurement scale; the mathematicians call this phenomenon fractals.

When the materials are in contact, the effective area, that means the surface where contact really exists, will be a fundamental parameter to be taken into account to

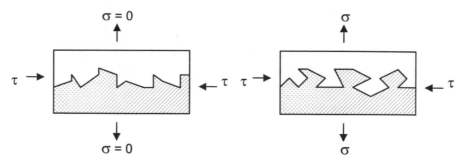

Fig. 4.4 Mechanical tensile and shear bond between substrate and overlay

explain the adhesion process. This is the result of the wetting procedure of the solid body by the liquid phase. The wetting procedure can be explained as follows [5]: the surface energies of the solid and the liquid interact each other and a change of the energy conditions occurs due to surface decrease of liquid/vapour and solid/vapour interfaces while a new interface (liquid/solid) is created. In this point of view, the contact angle is an interesting representation of this phenomenon: the lower the contact angle, the better is the spreading on the surface and the more effective will be the inter-molecular interactions at the interface. The third step is the development of these interactions – van der Waals forces – between the different phases. Their effect is based on the formation of electric fields of different intensities, depending on the presence of permanent or induced dipole bonds, or only dispersion bonds. They can be attributed to two different effects [8]:

1. *dispersion* forces arising from internal electron motions which are independent of dipole moments;
2. *polar* forces arising from the orientation of permanent electric dipoles and the induction effect of permanent dipoles on polarisable molecules.

Hydrogen bond forces can also be seen as a special case of dipole interactions: their range of actions extends further than that of the other secondary forces. At a higher level of energy, chemical bonds can occur when there is the development of covalent or ionic bond. This is the case if bonding agents are intentionally used, particularly at the interface polymer/mineral substrate (e.g., silane family products). Pigeon et al. [9] stated that the transition zone that is formed when new concrete is cast against old concrete is very similar to bond between aggregates and cement paste.

Mechanical adhesion in tension differs significantly from that in shear. For example, a high interface roughness usually improves the shear bond strength, whereas the tensile mechanical bond strength primarily depends on vertical anchorage in pores and voids (Figure 4.4). In practice, the difference is most obvious after debonding, if any.

4.4 Factors Affecting Bond

4.4.1 Concrete Properties

Materials for bonded overlays include:

- Portland cement mortar and concrete;
- Silica fume mortar and concrete;
- Polymer-modified mortar and concrete;
- Steel fibre reinforced concrete; and
- Pre-packaged repair mortar and concrete.

Of course, the selection of the overlay material influences the material properties of the overlay and, hence, also the bond. However, this chapter is focused on bond, whereas Chapter 2 is devoted to the overlay and the overlay materials. Consequently, this section is limited to a general discussion on the influence of fresh and hardened material properties on the bond.

4.4.1.1 Fresh Material Properties

The fresh repair material properties are very important both for early age bond strength development and bond durability. The workability and compaction of the freshly placed overlay influence the ability to fill open cavities and voids on the substrate concrete surface, thus determining the effective contact area between the two composites. Concrete of conventional fluidity and workability is generally sufficient in this respect. Small repair patches are commonly carried out with premixed, relatively stiff mortars, which are applied with a trowel. Self-consolidating materials (with high workability) are expected to lead to higher effective contact area and then to higher bond strength.

4.4.1.2 Hardened Material Properties

The compressive strength of the repair material usually does not influence the bond strength significantly. However, tensile strength is important as it affects crack development and, therefore, the formation of boundary conditions that may support the initiation of debonding. Delatte et al. [10] found that an increase in early age concrete strength increased both tensile and shear bond strength significantly.

Overlay permeability may influence bond durability, for example very impermeable overlays result in stresses at the interface when moisture from the substrate cannot migrate through the overlay [11].

The addition of polymers to cementitious repair mortars was found to result in better bond characteristics on specimens subjected to extensive temperature cycles [12]. Chen et al. [52] measured a significant increase in shear bond strength with the

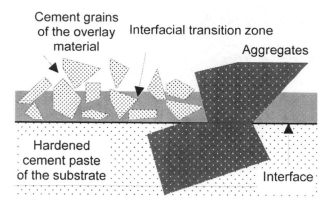

Fig. 4.5 Transition zone between substrate and overlay, according to Pigeon and Saucier [9]

addition of short carbon fibres to repair mortar. They attributed the effect to the decrease in drying shrinkage and the resulting decrease in interface stress. Granju [13] states that fibres enhance bond durability through the control of crack development.

Repair patch dimensions, such as area and thickness, can affect bond durability due to the influence on stresses resulting from differential movement between substrate and overlay. In general, large repairs tend to crack more easily than smaller areas. According to Banthia and Bindiganavile [14], thin repairs are more likely to debond than thicker repairs. On the contrary, Laurence et al. [15] have concluded from their measurements that bond strength was not influenced by the repair thickness. Equal bond strength does not necessarily mean that the likelihood of a bond failure is equal. The failure likelihood is dependent on both bond strength and bond stress, and the thickness is likely to influence the bond stress, e.g., by thickness dependent shrinkage stresses.

4.4.1.3 Overlay Strength

Pigeon and Saucier [9] state that the interface between old and new concrete is very similar to bond between aggregates and cement paste. According to their research, a wall effect exists between overlay and substrate, resulting in a transition zone that creates a layer of weakness (Figure 4.5).

Van Mier [16] has summarized existing knowledge on interfaces between aggregates and cement matrix. The bond mechanisms between aggregate and cement paste depend largely on the porosity of the aggregate. Generally a thin layer of calcium hydroxide (CH) forms at the physical boundary between aggregate and cement matrix, followed by a relatively open layer containing oriented CH crystals, ettringite, and calcium silicate hydrate (CSH). This so-called contact or transition layer has a high porosity. Van Mier explains this high porosity with absorption of mixing water at the surface of aggregate particles, which increases the effective w/c ratio. According to his research, fracture surfaces generally exist not directly at the physical boundary between aggregate and matrix but rather slightly remote from the interface in the porous transition zone.

These mechanisms have not yet fully been investigated in relation to interface between concretes of different ages but may be useful for the characterization of fundamental bond properties in composite members. Misra et al. [17] found a relation between air permeability at the interface and bond strength between substrate and overlay, which could be linked to the effects described above. Beushausen [18] found that "interface" failure in shear bond tests commonly occurs inside the overlay very close to the interface, which was observed both on short-term specimens and specimens tested after more than two years. Interface shear bond strength was closely related to compressive overlay strength, with the ratio between the two being time- and overlay strength-independent at approximately 0.10 to 0.12 (compare Section 4.4.11). This supports the theory that an interfacial transition zone exists as illustrated in Figure 4.5, representing a zone of weakness. On well-prepared substrates, overlay strength is therefore one of the decisive factors for bond strength and bond durability.

4.4.2 Removal of Deteriorated Concrete

The method of concrete removal has a major influence on both the surface of the remaining concrete and the properties of the uppermost layer of the remaining concrete. Some removal methods leave a rough and sound surface that promotes a good bond, while others may even introduce micro-cracks to a certain depth in the remaining concrete. Some removal methods can only remove a thin layer of concrete, while others have the ability to remove significant depth. Some of the most frequently used methods for removing concrete are summarised in Table 4.1. Additional methods, e.g., other hand-held tools, spring-action hammers, concrete crushers, drills and saws, non-explosive demolition agents, and jet-flame cutting are described by ACI Committee 555. According to Silfwerbrand [24], only water-jetting (or hydrodemolition) has the ability to remove concrete selectively, i.e., remove bad, damaged, and cracked concrete while leaving strong, sound, and undamaged concrete.

The surface of the remaining concrete needs to be free from microcracking; otherwise, the top layer of the remaining concrete will constitute a zone of weakness. The amount of microcracking is governed by the selected method of concrete removal. In principle, mechanical methods (hammers) are likely to introduce microcracking, whereas waterjetting has been shown to be a lenient – but efficient – method [24]. A comparison between surfaces treated by waterjetting and pneumatic hammers, respectively, is shown in Table 4.2. Because of microcracking, the bond strength was considerably lower in the case where the surface was treated with pneumatic hammers. Field tests, however, have shown that the bond strength can reach satisfactory values, if mechanical removal is followed by high-pressure water cleaning. The field tests also included the removal of a rutted top part of a concrete pavement by scarifying and placement of a new concrete overlay. Average

J. Silfwerbrand et al.

Table 4.1 Methods of concrete removal

Removal method	Principle behaviour	Concrete removal capability Action depth (mm)	Important advantages	Important disadvantages
Sandblasting	Blasting with sands.	No	No microcracking.	Not selective, leaves considerable sand.
Scrabbling	Pneumatically driven bits impact the surface.	Little (6)	No microcracking, no dust.	Not selective.
Shotblasting	Blasting with steel balls.	Little (12)	No microcracking, no dust.	Not selective.
Grinding (planning)	Grinding with rotating lamella.	Little (12)	Removes uneven parts.	Dust development, not selective.
Flame-cleaning	Thermal lance	No	Effective against pollutions and painting, useful in industrial and nuclear facilities.	The reinforcement may be damaged, smoke and gas development, safety considerations limit use, not selective.
Milling (scari-fying)	Longitudinal tracks are introduced by rotating metal lamellas.	Yes (75)	Suitable for large volume work, good bond if followed by water flushing.	Microcracking is likely, reinforcement may be damaged, dust development, noisy, not selective.
Pneumatic (jack) hammers (chipping), hand-held or boom-mounted	Compressed-air-operated chipping	Yes	Simple and flexible use, large ones are effective.	Microcracking, damages reinforcement, poor working environment, slow production rate, not selective.
Explosive blasting	Controlled blasting using small, densely spaced blasting charges.	Yes	Effective for large removal volumes.	Difficult to limit to solely damaged concrete, safety and environmental regulations limit use, not selective.
Water-jetting (hydro-demolition)	High pressure water jet from a unit with a movable nozzle	Yes	Effective (especially on horizontal surfaces), selective, does not damage reinforcement or concrete, improved working environment.	Water handling, removal in frost degrees, costs for establishment.

bond strength of 2.3 MPa was obtained [19]. Talbot et al. [20] and Carter et al. [21] found that sandblasting subsequent to the use of heavy mechanical methods could remove the damaged concrete and provide a sound interface. Wells et al. [22] and Warner et al. [23] achieved good bond strength on surfaces that were sandblasted without prior roughening.

Table 4.2 Laboratory pull-off tests: influence of microcracking [24]

Interface treatment	Presence of microcracking	All tests		Interface failures	
		Number of cores	Average failure stress (MPa)	Number of cores	Average failure stress (MPa)
Waterjetting	No	16	1.86	1	2.23
Pneumatic hammers	Yes	16	1.10	5	0.94

4.4.3 Concrete Removal behind Rebars, Rebar Cleaning, and Rebar Replacement

According to Silfwerbrand [25], the interface between substrate and shotcrete should not coincide with the reinforcement plane because it may affect bond durability under cyclic loading. According to the bridge maintenance regulations of the Swedish National Road Administration [26], the entire reinforcement bar has to be completely surrounded by new concrete if more than 30% of the circumference has been exposed through concrete removal operations. The distance between bar and substrate has to be equal to the maximum aggregate size of the overlay concrete plus 5 mm. Vaysburd et al. [27] give a similar recommendation:

> concrete removal around the bar shall continue to provide a minimum 19 mm clear space between the rebar and the surrounding concrete or 6 mm larger than the maximum size aggregate in the repair material.

Vaysburd et al. also discuss other reasons than bond for exposing the rebar. The most obvious one is to clean the rebar from corrosion products. The other is to remove the concrete surrounding the corroded rebar, concrete that can be highly stressed due to the corrosion products' volume increase.

Exposed reinforcement bars should be thoroughly cleaned from dust and corrosion products. The Swedish National Road Administration [26] states that corroded reinforcement shall be replaced by new with equal area. New bars thoroughly anchored should replace reinforcement bars that loosen or have lost bond to the substrate concrete.

4.4.4 Cleaning after Concrete Removal

The single most important factor influencing bond is the surface cleanliness. A surface which is contaminated at the time of overlay placement will produce poor bond characteristics. In the first Swedish bridges repaired with waterjetting and bonded overlays in 1984 and 1985, coring showed poor bond at several locations. In most cases, insufficient cleaning was the reason for this. Loose particles were found in the interface between old and new concrete [24]. In order to warrant cleanliness, the wa-

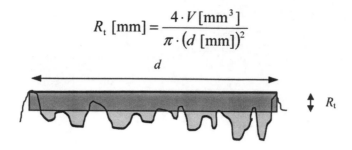

$$R_t \, [\text{mm}] = \frac{4 \cdot V \, [\text{mm}^3]}{\pi \cdot (d \, [\text{mm}])^2}$$

Fig. 4.6 Determination of surface roughness using the sand area method [28]

terjetted surface needs to be cleaned twice. The first cleaning has to be done shortly after waterjetting to prevent loose concrete particles, such as exposed un-hydrated cement surfaces, to bond to the surface. The second cleaning has to be carried out prior to overlay placement to make sure that the surface is free from sand, oil, dust or other particles with origin from the environment or the construction works. The best methods are hosing down with high-pressure water and the use of vacuum cleaners.

4.4.5 Surface Properties

Interface roughness depends to a large extent on the method of substrate surface preparation. Mechanical methods of concrete removal normally leave the substrate surface much rougher than blast methods. The magnitude of surface roughness for concrete repairs is commonly measured in mm. The most widespread test method is the sand area method (Figure 4.6) [28], in which sand of known volume V is spread over the concrete surface to form a circle until all sand has settled in the surface cavities. The roughness R_t can be calculated from the diameter d of the circle, using the following equation:

$$R_t \, [\text{mm}] = \frac{4 \cdot V \, [\text{mm}^3]}{\pi \cdot (d \, [\text{mm}])^2} \tag{4.1}$$

Other methods are used in Belgium and by the International Concrete Repair Institute (ICRI). Still another surface roughness method using a saw-tooth curve (Figure 4.7) was introduced by Silfwerbrand [29]. More accurate surface profiles can be measured electronically using touch-pins [7] or laser techniques [30].

The surface roughness used to be considered to have a major influence on the bond between old and new concrete. Bond tests have, however, shown that surface roughness only has a minor influence on the bond (Table 4.3). In a test series, bond to rough, waterjetted surfaces was compared with bond to smooth, sandblasted surfaces [24]. The average bond strength was approximately equal, but interface failures were more frequent on the sandblasted surface. The conclusion is

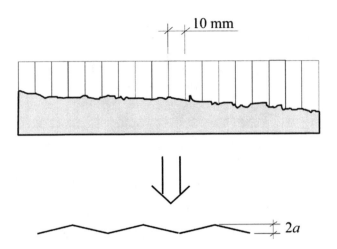

Fig. 4.7 The surface profile is transferred to a saw-tooth curve with double amplitude $2a$. The distance between consecutive measuring points is 10 mm [24, 29]

Table 4.3 Laboratory pull-off tests: influence of surface roughness [24]

Interface treatment	Surface roughness*	All tests		Interface failures	
	Double amplitude (mm)	Number of cores	Average failure stress (MPa)	Number of cores	Average failure stress (MPa)
Waterjetting	7.7	16	1.86	1	2.23
Sandblasting	0.4	8	2.38	3	1.73

* Surface roughness was measured on surface profiles transferred to a saw-tooth curve with two characterising parameters: double amplitude and wavelength (see Figure 4.7).

that there might be a threshold value. If the surface roughness is higher than the threshold value, further improvement of the roughness does not seem to enhance bond strength. According to these tests, this threshold value ought to be close to the surface roughness of typical sandblasted surfaces. However, further investigations using interface shear bond tests are recommended for the verification of this theory for different bond mechanisms.

The existence of a threshold value is evidenced by tests by Takuwa et al. [31]. They studied the influence of surface preparation of the substrate concrete and defined "increase of area" as a surface roughness parameter. They compared manual treatment, shot blasting treatment and water-jetting with no treatment (Figure 4.8).

Mainz and Zilch [32] achieved high bond strengths on water-jetted surfaces with a roughness of >1 mm. Similarly, Tschegg et al. [33] compared roughness of 1.75 to 0.65 mm on water-jetted surfaces and found better bond characteristics for the rougher interface.

The laitance is a layer of weak and non-durable material containing cement and fines from the aggregates, brought by bleed water to the top of the concrete. If the

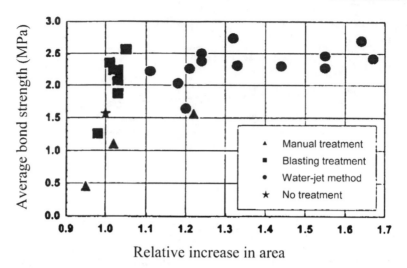

Fig. 4.8 The relationship between the increase of surface roughness (described as increase of substrate area in percent) and bond strength [31]

laitance is not removed prior to overlay placement, it will lower the bond markedly. If the new overlay replaces old concrete, the laitance will be removed together with the deteriorated concrete. However, if the overlay is placed on unaltered parts, the removal of the laitance must not be forgotten. Usually, sandblasting will be sufficient to remove the laitance.

The old weathered and carbonated concrete surface can be removed using water-jetting, thus exposing a fresh and uncontaminated surface. However, long periods of time between substrate resmoval and overlay placement may result in new carbonation of the exposed surface, especially in dry environments. Test results by Gulyas et al. [34] show that carbonation may decrease the bond strength significantly. Block and Porth [35], on the contrary, found that substrate carbonation does not affect pull-off bond strength.

The substrate temperature at the time of overlay placing was found to have a significant effect on shear bond strength development. Cold substrate (4°C) results in lower initial bond strength but higher long-term bond strength, compared to substrates of higher temperature (21 or 38°C) [36].

One line in two perpendicular directions of the treated surface was measured every 0.1 mm using a laser device. The area increase was determined by transferring the obtained zigzag lines. Note that an area increase also was measured for the non-treated surface (indicated with a star).

4.4.6 Surface Preparation

According to Vaysburd et al. [37], the "process of concrete surface preparation for repair is the process by which sound, clean and suitably roughened surface is produced on concrete substrates". The workmanship is of uppermost importance. Besides removal of unsound concrete and all foreign materials that may disturb bond development, the process also covers the opening of the substrate concrete pore structure.

The influence of surface moisture on the bond between old and new concrete has been investigated in many studies. A too dry surface may absorb water from the fresh concrete overlay giving risk of a heterogeneous, porous zone close to the interface. If the surface is too wet, an overlay zone with a high water-cement ratio will develop close to the interface, which will lead to a local reduction in overlay strength. Free water at the surface may destroy the bond completely. Zhu [38, 39] has found experimental signs of optimal moisture, but the moisture influence on the bond was so small that it was difficult to discern between moisture influence and scatter of test results. Li et al. [40] measured the bond strength of repaired specimens after freezing and thawing cycles and found that different repair materials correspond to different optimum interface moisture conditions at the time of casting. There is widespread agreement to promote the "saturated substrate with dry surface" as one of the best compromises [41].

4.4.7 Bonding Agents

Bonding agents, e.g., Portland cement grout, latex modified Portland cement grout, and epoxy resins, are sometimes used to improve bond [42]. However, bonding agents cannot compensate for bad substrate surface preparation and may act as a bond breaker when used inappropriately [9, 11]. The chapter author's opinion is that bonding agents should normally be avoided. The use of bonding agents leads to two interfaces and thus to the creation of two possible planes of weakness instead of one. Besides, a grout often has a high water-cement ratio leading to a low strength and the risk of a cohesive failure within the bonding agent itself. On the other hand, grouts may have an ability to assimilate loose particles on an insufficiently cleaned surface. This assimilation may increase the bond strength for this specific case. Bonding agents may improve bond strength for certain materials, especially stiff repair mortars that cannot properly fill open pores and cavities.

4.4.8 Mechanical Devices Crossing the Interface

In cases of poor or uncertain bond or in areas with specific demands (e.g., high shear or tensile stresses, serious consequences in the case of failure), it might be

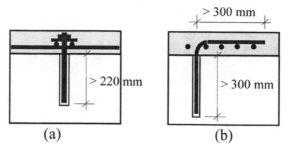

Fig. 4.9 Example of dowel bars crossing the interface. (a) Nut, plate, and stud M20 in injected hole $\Phi = 30$ mm. (b) Bent reinforcement bar $\Phi = 16$ mm in injected hole $\Phi = 25$ mm

necessary to strengthen the shear and tension capacity by installing some kind of reinforcement crossing the interface. It is mandatory that the reinforcement be sufficiently anchored in both remaining concrete and in the overlay. Two Swedish solutions are given in Figure 4.9. It is important to note that the reinforcement units do not work until the bond has broken. The reason is that the reinforcement unit has to be strained before it carries more than a negligible part of the load. This is similar to ordinary reinforcement that mainly carries load after concrete cracking.

4.4.9 Concrete Placement

The compaction is important to obtain a dense and homogeneous overlay as well as a good and uniform bond. Especially, the compaction is important in overlays on rough surfaces to prevent the development of air pockets in the valleys of the surface texture. Air pockets were found in some cores taken from repairs in the early eighties [24]. The Swedish National Road Administration [26] recommends the use of vibration pokers and vibration platforms.

4.4.10 Concrete Curing

Early age shrinkage may result in overlay cracking. Cracks may initiate debonding due to the formation of boundary conditions (free edges). Curing is therefore one of the most important factors in reducing early-age overlay and interface stresses. It prevents moisture loss and thereby reduces early age shrinkage, leading to higher tensile strength at the onset of shrinkage (Figure 4.10). Simultaneously, other advantages are gained: reduced risk of plastic cracking, higher strength, improved durability, and better wear resistance. Paulsson and Silfwerbrand [25] recommend a minimum of five days water curing on bridge deck overlays. Exposure to direct sunlight was found to have a detrimental effect on shear bond strength even under

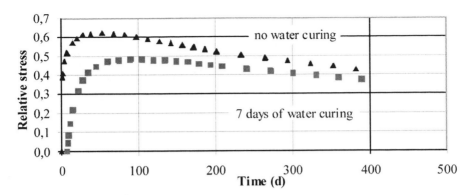

Fig. 4.10 Relationship between relative stress (equals ratio between maximum tensile stress and available tensile strength) and time after casting – computational result according to Silfwerbrand [29]

proper curing conditions which included wet burlap and plastic [10]. Li et al. [40] tested different repair materials and found that, for some materials, wet curing, and for others, dry curing results in better bond strength. Schrader [11], on the contrary, states that curing mainly affects the surface of a concrete repair, but has little influence on the material or bond properties at a depth of more than 25 mm.

4.4.11 Short-term Bond Properties

The development of early age bond strength is important for the structure's ability to withstand interface stresses induced by early differential movement between substrate and overlay. For pavement and bridge deck overlays high early bond strength is usually required due to traffic and live loads. According to Delatte et al. [10, 36], bond strength develops rapidly after placement, similar to concrete compressive strength development. In their studies they suggest a concrete maturity approach, which characterises bond strength development in relation to the concrete's rate of hydration rather than its age. Laboratory tests carried out by Silfwerbrand [42] have shown that the bond strength development is rapid (Figure 4.11). At an age of 24 hours, the obtained bond strength exceeded 1 MPa. The bond strength development was found to be faster than the development of compressive strength of the concrete overlay. The tests were carried out indoors. At time of testing, the old concrete had compressive cube strength of 54 MPa. The overlay had 28-day compressive cube strength of 63 MPa. The surface of the old concrete was waterjetted prior to overlay placement.

Beushausen [18] measured the development of bond strength with time, using a modified version of the direct shear test proposed by FIP [43] (compare Section 4.5). Test specimen size was 150 × 150 × 50 mm with both substrate and overlay having a depth of 75 mm. The interface was sandblasted with an average roughness of

J. Silfwerbrand et al.

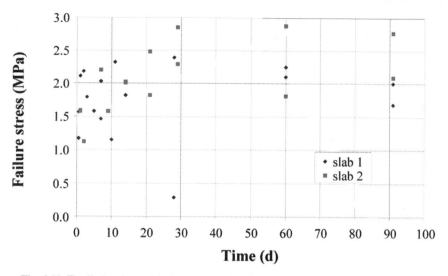

Fig. 4.11 Tensile bond strength development according to Silfwerbrand's test results [42]

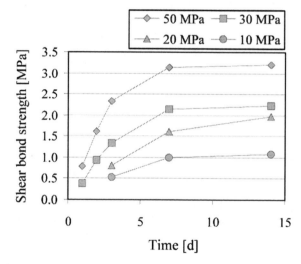

Fig. 4.12 Interface shear bond strength development for overlays of different compressive strengths

0.7 mm. All specimens were fully water cured at 23°C. Different overlay strengths were used (28-day design strengths 10, 20, 30, and 50 MPa), the substrate compressive strength being 48 MPa. Shear bond strength development was closely related to the development of compressive overlay strength, with a relatively constant ratio between the two of approximately 0.10 to 0.12 over the whole test period for all specimens. Shear bond strength development is shown in Figure 4.12. The test results represent the mean value of 4 to 6 specimens after exclusion of outliers.

For fully bonded overlays it appears that the rate of bond strength development can be related to that of mechanical overlay strength. This facilitates the design of interface shear strength for structural overlays.

Carter et al. [21] state that bond strength develops more fully in the centre of an overlay as the boundaries are especially subjected to cyclic stresses related to differential temperature and moisture content.

4.4.12 Long-term Bond Properties

Differential movement between substrate and overlay, such as differential shrinkage and thermal expansion and contraction, is commonly the main factor affecting long-term bond properties. It ought to be emphasized that high initial bond strength does not necessarily lead to bond durability. Proper substrate preparation, material selection and curing procedures should, however, normally result in good long-term bond properties. Repaired beams and columns with well-bonded overlays were shown to have structural capacities similar to that of monolithic members [29, 44]. Carter et al. [21] state that well designed bridge deck overlays can be expected to provide more than 30 years of service life if they are placed and cured correctly. Paulsson and Silfwerbrand [45] tested repaired concrete bridge decks, which had been overlaid with steel fibre reinforced concrete and found a slight increase in bond strength after a period of 9 years. Specimens subjected to 300 freezing and thawing cycles were found to have similar bond strength as air cured specimens [40]. Talbot et al. [20] investigated the influence of different interface textures and concluded that smooth surfaces as well as sandblasted surfaces experienced a significant loss of bond strength with time. However, surfaces which were roughened mechanically and subsequently sandblasted had good bond durability. Debonding on vertical or overhead concrete repair patches usually leads to spalling and hence to failure of the repair. Delaminated bridge deck overlays, on the contrary, may be re-bonded using epoxy injection [46].

4.4.13 Traffic Vibrations

Normally, traffic-induced vibrations do not harm the development of bond between old and fresh concrete. Independent researchers have even found that continuous and limited vibrations may increase both overlay strength and bond strength, see, e.g., Silfwerbrand [42]. Heavy vibrations starting a few hours after overlay placement should, however, be avoided. Manning [47] stated that the best way of preventing heavy vibrations is to maintain a smooth riding surface and a smooth transition at the expansion joints of the bridge.

4.5 Test Methods

4.5.1 General

The results and interpretation of interface bond tests depend to a large extent on the test method used. Common bond test methods include interface shear, torsion and tension tests and a wide range of possible test set-ups have been developed for laboratory testing. Interface shear strength values obtained by different test methods may differ substantially as test results depend on specimen size, test set-up, loading rate, etc. Li et al. [40] investigated the size effect in bond tests and concluded that smaller specimen sizes led to larger bond strength in prism splitting tests. A comparison of test results obtained with different test methods, therefore, is problematic.

Most cracks in concrete develop under combined modes, of which the mixed mode 1 (tension) and 2 (in-plane shear) is the most common. According to Van Mier et al. [48], most interface shear tests will initiate in mode 1, and true mode 2 failures are very difficult to obtain. Robins and Austin [49] proposed a bond failure envelope for concrete repairs based on the Mohr circle and related it to interface shear and tension test results.

Many test methods have been developed to determine interface bond strength between concrete substrate and overlay (Figure 4.13).

In a direct tension test, the specimen is pulled apart by loads applied perpendicular to the bonded interface. The most common test method is the pull-off test, which can be carried out in-situ (Figure 4.13A). The pull-off test encounters several problems as the test results may be strongly influenced by eccentricity in the load application and damage during coring [36]. Laboratory tensile strength tests (Figure 4.13B) are of course possible. They give especially reliable results when the test is performed with the lowest possible eccentricity and no hinge in the loading chain. After debonding initiation at the weaker point of the interface, hinges allow debonding propagation by flexure, resulting in lower measured strength and higher scatter of the results.

Tensile bond strength is hard to quantify on properly placed overlays, yet easy to quantify in cases where the tensile interface strength is less than the tensile material strength. The results of pull-off tests, therefore, often only give indication of the lower bound of the interface bond strength. The main stress situation at the interface between substrate and overlay is that of a mixed mode of tension and shear (compare Section 2.3) and bond characterisation by sole tensile strength seems problematic. Weber [50] indicates the usefulness of pull-off bond tests at locations where tensile interface stresses are predominant, for example at the boundaries of a repaired member. Vaysburd and McDonald [37] have evaluated the pull-off test method through 257 cores on 77 experimental repairs. Their main results can be summarized in the following conclusions:

- Results obtained from pull-off tests were variable.
- Variation in exposure conditions did not appear to have a significant effect on failure mode or bond strength.

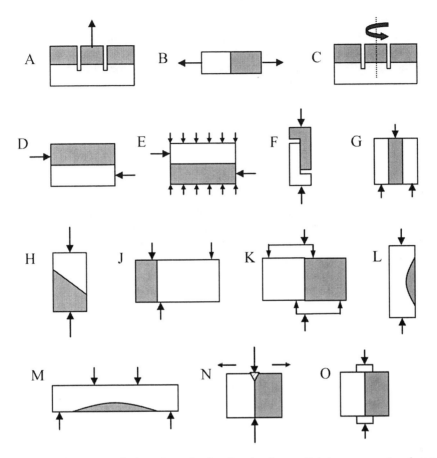

Fig. 4.13 Various test methods to determine interface bond strength between concrete substrate and overlay

- Only two of three pull-off test devices investigated ought to be used to evaluate bond strength.
- The load must be perpendicular to the repair surface, attached centrally to the core and increased gradually with a specified speed (Vaysburd and McDonald [37] recommend 0.02–0.05 MPa/s, Silfwerbrand [29] has used 0.064 MPa/s).
- The depth of the core-drilling into the substrate below the interface should be a minimum of 25 mm or one-half of the core diameter, whichever is larger.

Silfwerbrand [2] developed an in-situ test for torsional bond strength, which can be equated to interface shear strength (Figure 4.13C). Common shear bond tests apply a force parallel to the interface (Figure 4.13D). Modified shear bond methods have been presented by Pigeon and Saucier [9] (Figure 4.13E) and El-Rakib et al. [51] (Figure 4.13F). The push-out specimen [52] (Figure 4.13G) has the disadvantage of having two interfaces, which does not represent site conditions, thus making the test very impractical The slant shear test (BS 6319) (Figure 4.13H) measures bond strength under a combination of shear and compression. Several researchers [10, 53, 54] indicated shortcomings of this test with respect to unrealistic loading conditions and the vast number of parameters which may affect the test results. The guillotine test (Figure 4.13J) can be used both on cores and prisms. Delatte et al. [10], however, reported difficulties with the alignment of the loading head for use on cores.

4.5.2 Shear Test Methods

The disadvantage of most common shear test methods is the occurrence of an interface bending moment due to force eccentricity. The FIP [43] developed a test method for interface shear strength, in which the interface is theoretically subjected to pure shear forces (Figure 4.13K). Robins and Austin [49] developed different patch tests, which measure interface shear and tensile bond strength under structural loading (Figures 4.13L and 4.13M). The wedge splitting test device [33] characterises bond by fracture mechanical parameters such as crack opening and specific fracture energy, and tensile interface strength in bending (Figure 4.13N). Li et al. [40] measured interface tensile strength with a prism splitting test device (Figure 4.13O).

In cases where destructive in-situ bond testing is not appropriate, chain dragging or tapping the surface with a hammer may be used to determine locations of debonding [21]. Lacombe et al. [55] analysed bond characteristics of different repair materials visually and by a scanning electron microscope. They evaluated the quality of the bond by observation of overlay compaction, cracks and voids at the interface.

All shear test methods described above have the disadvantage in common that the test specimens have to be prepared in the laboratory. In some cases, the specimens even have to be cast in the laboratory. Consequently, a test method that can be conducted in-situ is desired. The Department of Structural Engineering, KTH, has developed such a method [2, 29]. The test specimen is of the same dimensions as the one used in-situ for pull off test. The difference is that a torsional moment is applied to the core instead of a tensile force (Figure 4.13C).

The apparatus for determining the shear bond strength consists of a stand, three steel bars, a moment converter (25 times amplification), a torsion gauge, rotational and spherical bearings, a cylindrical steel plate, and a data acquisition unit holding the maximum value (Figure 4.14). The stand is fixed to the concrete surface with three expander bolts. The steel plate is glued to the top surface of the drilled core that is still attached to the old concrete. The three bars serve as a moment converter. The bearings prevent the development of normal forces and shear forces, and are fixed to three holes in the steel plate. A torsional moment is applied to the top of the bars by using the bending moments. During the test, a torsional moment T is successively increased until failure. Assuming linear elastic behaviour, the relationship between shear stress τ and torsional moment can be computed by

$$\tau = \frac{16}{\pi} \cdot \frac{T}{\varphi^3} \qquad (4.2)$$

where φ is the core diameter. For purely plastic materials, the relationship between shear stress and torsional moment has the following expression:

$$\tau = \frac{12}{\pi} \cdot \frac{T}{\varphi^3} \qquad (4.3)$$

i.e., for a constant torsional moment, the shear stress for a plastic material is 25% less than corresponding value for an elastic value. Materials having other constitutive laws ought to have shear stress–torsional moment relationships somewhere between these two extremes. Thus, the error using either of the two expressions should be limited. Concrete is a brittle material and, consequently, the linear elastic relationship (Equation (4.2)) has been applied. Numerical analyses of the stresses and strains within the test specimen would be desirable, but have not been included in this investigation.

The shear stress τ equals the shear strength τ_{max} when the torsional moment T reaches the failure moment T_{max}. Only in cases with complete interface failures, the shear strength is equal to the shear bond strength. For other failure modes, the shear strength constitutes a lower bound of the shear bond strength. At KTH, the core diameter has been 100 mm and the stress velocity approximately 0.1 MPa/s.

Possible sources of error in addition to the one due to simplifications when transferring torsional moment to shear strength are normal forces that may occur simultaneously when applying the torsional moment and calibration errors when transferring the gauge signal to a moment value. As mentioned above, the bearings have been designed to minimize the influence of normal forces. A rough estimation limits the sum of these latter errors to 10% of the measured value. This is of about the same magnitude as for pull-off tests measured in-situ. It should be noted that shear and tensile strengths of concrete both have a rather large inherited scatter. In order to use this test method for standardised laboratory tests, further studies and further development of the test apparatus are needed.

4.5.3 Comparisons between Tensile Bond Strength and Shear Bond Strength

Relating interface shear and tension tests is questionable as both bond mechanisms have substantially different characteristics (compare Section 2.2.1). However, Silfwerbrand [2] and Delatte et al. [10] indicated a correlation between the two test methods. The latter measured a mean ratio (shear bond divided by tension bond) of 2.04.

Measurements show that the shear bond strength is considerably higher than the tensile bond strength. Analysing the test results available [2], the ratio between average torsional shear stress and average tensile stress varies between 1.9 and 3.1.

Fig. 4.14 Torsion test apparatus [29]

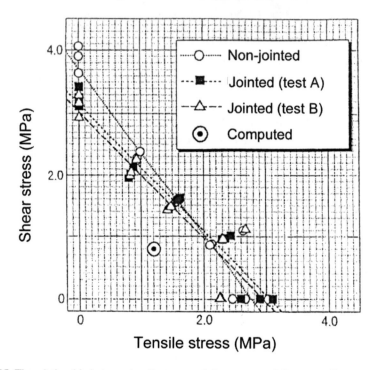

Fig. 4.15 The relationship between tensile stress and shear stress at failure according to test results by Sato [56]

For water-jetted surfaces, the obtained shear bond strength exceeds 3 MPa. In Japan, Sato [56] has found a relationship between shear strength and tensile strength slightly lower than 1.5 by tests on specimens subjected to solely tension, solely shear or combinations of tension and shear (Figure 4.15).

Table 4.4 Bridges investigated [25]

Bridge	Structural system	Climate zone	Year of construction	Year of bridge deck repair	Average daily traffic	Percentage of heavy vehicles	Use of de-icing salt
Bjurholm	Steel girders	5	1966	1985	1 440	12	occasional
Mälsund	Concrete arch	2	1924	1986	210	10	occasional
Skellefteå	Steel girders	4	1960	1987	25 000	8	moderate to intensive
Södertälje	Flat slab	2	1966	1989	~10 000	10	intensive
Umeå	Steel girders	4	1949	1987	23 000	20	moderate to intensive
Vrena	Steel girders	2	1937	1987	170	10-15	occasional
Överboda	Concrete arch	2	1946	1986	7 600	10	intensive

Note: Climate zone 2 means an annually frost amount of 300–600 day×degree of frost, zone 4: 900–1200, and zone 5: 1200–1500 (300 day×degree of frost = 30 days of –10°C or 60 days of –5 °C).

4.6 Evaluation

In the mid 1980s, more than 250 Swedish concrete bridge decks were repaired by means of water-jet removal and a new-cast, bonded concrete overlay. A few weeks after repair, the Department of Structural Engineering, Royal Institute of Technology, tested the bond on more than 20 bridges [24, 29]. The average failure tensile stress was above 1.5 MPa. The bond strength might have been even higher, since only a small percentage of the tested cores failed at the interface. If the failure occurs elsewhere, the failure stress only gives a lower bound of the bond strength.

Between 1995 and 1999, a major research project was carried out at the Department of Structural Engineering to study durable repairs [25, 45, 57, 58]. Seven bridges repaired in the mid 1980s were selected. Tests on concrete strength, bond strength, and freeze-thaw resistance were carried out as well as crack surveys and measurements on chloride penetration.

In previous sections, the most important factors influencing bond have been discussed. They are connected to measures taken prior to, during, and shortly after overlay placement. The long-term performance of the bond may, however, also be influenced by traffic loads and environmental factors. The following factors were considered to be most important: (i) the number of heavy vehicles, (ii) the stiffness of the bridge, in turn dependent on the structural system, (iii) the use of de-icing salts, and (iv) the number of freeze-thaw cycles. Within the bridge, the bond may be different in zones with positive moment and negative moment, respectively. The seven bridges chosen have different traffic volumes, different structural systems, locations in different climate zones, and different needs for de-icing salts (Table 4.4). Consequently, all factors were covered in the investigation, even though all could not be studied separately.

Table 4.5 Field tests on bond at time of repair and ten years afterwards [25]

Bridge	Number of cores tested		Failure stress (MPa)		Failure stress increase (%)
	at time of repair	in 1995	at time of repair	in 1995	
Bjurholm	9	8	1.96	1.99	2
Mälsund	6	8	1.71	2.17	27
Skellefteå	3	9	1.43	1.82	27
Södertälje	?	7	> 1.5	2.05	~ 35
Umeå	4	9	1.56	1.61	3
Vrena	8	9	1.49	1.56	5
Överboda	12	9	2.09	2.18	3

Bond strength was measured on cores drilled from the concrete bridge decks. The cores had a nominal diameter of 100 mm. At time of repair, the pull-off tests were carried out in situ. In 1995, the pull-off tests were carried out in the laboratory some time after drilling. Between drilling and testing, the cores were stored in sealed plastic bags. Before testing, the cores were sawn perpendicularly to the cylinder axis. Steel plates were glued to the end surfaces. The cores were placed in a testing machine. A loading rate of 500 N/s (0.064 MPa/s) was chosen. Failure stresses and failure modes were registered. Most failures occurred in the old concrete. All tests are described in the test report [25, 57]. The results are summarised in Table 4.5.

On all bridge decks investigated, the failure stress had increased slightly during the time after repair. That means, that neither traffic loading, nor de-icing salts and frost-thaw cycles had deteriorated the bond between old and new concrete. The conclusion is that a good and durable bond can be obtained providing all works with concrete removal, surface cleaning and preparation, overlay placement, and curing are carried out correctly and carefully.

4.7 Design Strength Values

Several codes deal with shear resistance at the interface between two concretes. Usually, the design shear resistance consists of three parts: (i) the strength of the interface itself, (ii) a term dependent on external compression forces across the interface, and (iii) a term dependent on the reinforcement crossing the interface. In the following comparison, only the first part has been included. It should be noted that higher shear resistance values may be used when verified by extensive testing.

According to the Swedish handbook for concrete structures [59], the design shear resistance at a waterjetted, properly cleaned interface is equal to 0.4 MPa.

The American code ACI 318-99 [60] deals with composite structures in its chapter 17. Waterjetted surfaces are not dealt with. For clean surfaces, free of laitance, and intentionally roughened, the design shear resistance is limited to 0.55 MPa.

According to CEB-FIP MC 90 [61], the design shear resistance at the interface is proportional to the compressive strength f_{ccd}, if the surface is rough or indented. The coefficient of proportionality is equal to 0.06. Hence, if $f_{ccd} = 25$ MPa, e.g., the design shear resistance is 1.5 MPa. The code does not cover waterjetted surfaces.

The draft of Eurocode 2 [62] states that the design shear resistance at the interface is proportional to the design tensile strength f_{ctd} of the weakest concrete. The coefficient of proportionality is 0.45 for rough surfaces and 0.5 for indented surfaces. If $f_{ctd} = 1.6$ MPa, e.g., the design shear strength is 0.72 MPa at a rough interface. The draft does not deal with waterjetted surfaces.

Comparisons between the design strength and the shear strength obtained for waterjetted surfaces show a very large discrepancy. The code writers ought to discern between waterjetted and other rough surfaces and increase the design shear strength for waterjetted interfaces considerably. Another, maybe easier, way to deal with the problem is that the designer only considers shear stresses due to shear forces and ignores shear stresses due to differential shrinkage. Most of the latter stresses will disappear with time due to creep, and the likelihood that maximum shear force and maximum differential shrinkage stress occur simultaneously is very low. The maximum shear due to traditional shear forces (e.g., due to traffic, wind or snow) appears only once during the lifespan of 60 to 120 years, while shear stresses due to differential shrinkage peaks after some months. By comparing stresses caused by only one type of loading (e.g., traffic) with a considerably reduced strength, the result will still be far on the conservative side.

4.8 Performance Requirements

Standard EN 1504-3 [63] states that the tensile bond strength should equal or exceed 2.0 MPa on structural repairs and 1.0 MPa on non-structural ones. Considering obtained in-situ measurements and the fact that all deteriorated old concrete usually cannot be removed, these requirements may be fairly difficult to fulfil. The Swedish National Road Administration with a long experience on concrete bridge repair, has used and is still using [64] the following requirements:

$$m \geq f_v + 1.4 \cdot s \qquad (4.4)$$

$$x \geq 0.8 \cdot f_v \qquad (4.5)$$

where f_v is the required tensile bond strength equalling 1.0 MPa, m and s are the average and the standard deviation ($s = 0.36$ MPa) of the measuring values, respectively, and x is a single measuring value.

References

1. Courard, L., How to analyse thermodynamic properties of solids and liquids in relation with adhesion? In *Proceedings 2nd International RILEM Symposium ISAP99*, pp. 9–19, 1999.
2. Silfwerbrand, J., Shear bond strength in repaired concrete structures. *Materials & Structures*, **36**, 419–424, July 2003.
3. Derjagin, B.V., Krotova, N.A., and Smilga, V.P., *Adhesion of Solids*. Studies in Soviet Science: Physical Sciences, Plenum Publishing Corporation, New York, 455 pp., 1978.
4. Courard, L., Parametric study for the creation of the interface between concrete and repair products, *Materials and Structures*, **33**, 65–72, January–February 2000.
5. Fiebrich, M.H., Influence of the surface roughness on the adhesion between concrete and gunite mortars overlays. In *Proceedings 2nd Bolomey Workshop, Adherence of Young and Old Concrete*, Unterengstringen, Switzerland, F. Wittmann (Ed.), Aedification Verlag , pp. 107-114, 1994.
6. Kinloch, A.J., *Adhesion and Adhesives: Science and Technology*. Chapman and Hall, London, 441 pp., 1987.
7. Courard, L., Parametric definition of sandblasted and polished concrete surfaces. In *Proceedings IX ICPIC 98*, Bologna, Italy, September 14–18, pp. 771–778, 1998.
8. Fiebrich, M., Grundlagen der Adhäsionskunde, *Deutscher Ausschuss für Stahlbeton*, **334**, 75-90, 1994.
9. Pigeon, M. and Saucier, F., Durability of repaired concrete structures. In *Proceedings, International Symposium on Advances in Concrete Technology*, Athens, 11–12 May, pp. 741–773, 1992.
10. Delatte, N.J., Williamson, M.S., and Fowler, D.W., Bond strength development of high-early-strength bonded concrete overlays, *ACI Materials Journal*, **97-M27**, 201–207, March–April 2000.
11. Schrader, E.K., Mistakes, misconceptions, and controversial issues concerning concrete and concrete repairs, Parts 1, 2, and 3, *Concrete International*, September, October, and November 1992.
12. Atzeni, C., Massidda, L., and Sanna, U., Dimensional variations, capillary absorption and freeze-thaw resistance of repair mortars admixed with polymers, *Cement and Concrete Research*, **23**, 301–308, 1993.
13. Granju, J.L., Thin bonded overlays – About the role of fibre reinforcement on the limitation of their debonding, *Advanced Cement Based Materials*, **4**, 21–27, 1996.
14. Bindiganavile, V. and Banthia, N., Repairing with hybrid-fiber-reinforced concrete, *Concrete International*, 29–32, June 2001.
15. Laurence, O., Bissonnette, B., Pigeon, M., and Rossi, P., Effect of steel macro fibres on cracking of thin concrete repairs. In *Proceedings, 5th International RILEM Symposium on Fibre-Reinforced Concretes (BEFIB 2000)*, Lyon, France, pp. 213–222, 2000.
16. Van Mier, J.G.M., *Fracture Processes of Concrete*, CRC Press, 1997.
17. Misra, A., Cleland, D.J., and Basheer, P.A.M., Effect of different substrate and overlay concretes on bond strength and interfacial permeability, *Concrete Science and Engineering*, **3**(10), 73–77, 2001.
18. Beushausen, H., Long-term performance of bonded concrete overlays subjected to differential shrinkage. Dissertation, University of Cape Town, unpublished, 2005.
19. Silfwerbrand, J. and Petersson, Ö., Thin concrete inlays on old concrete roads. In *Proceedings, 5th International Conference on Concrete Pavement Design & Rehabilitation*, Purdue University, West Lafayette, Indiana, USA, Vol. 2, pp. 255–260, April 1993.
20. Talbot, C., Pigeon, M., Beauprè, and Morgan, D.R., Influence of surface preparation on long-term bonding of shotcrete, *ACI Materials Journal*, **91**(6), 560–566, November–December 1994.
21. Carter, P., Gurjar, S., and Wong, J., Debonding of highway bridge deck overlays, *Concrete International*, 51–58, July 2002.

22. Wells, J.A., Stark, R.D., and Polyzois, D., Getting better bond in concrete overlays, *Concrete International*, 49–52, March 1999.
23. Warner, J., Bhuyan, S., Smoak, W.G., Hindo, K.R., and Sprinkel, M.M., Surface preparation for overlays, *Concrete International*, 43–46, May 1998.
24. Silfwerbrand, J., Improving concrete bond in repaired bridge decks, *Concrete International*, **12**(9), 61–66, September 1990.
25. Silfwerbrand, J. and Paulsson, J., The Swedish experience: Better bonding of bridge deck overlays, *Concrete International*, **20**(10), 56–61, 1998.
26. Swedish National Road Administration, Technical Regulations for Bridges Maintenance, Borlänge, Sweden, 90 pp., 2002 [in Swedish].
27. Vaysburd, A.M., Sabnis, G.M., and McDonald, J.E., Interfacial bond and surface preparation in concrete repair. *ICJ*, **75**(1), 27–33, 2001.
28. Kaufmann, N., Das Sandflächenverfahren, *Strassenbau Technik*, **24**(3), 31–50, 1971.
29. Silfwerbrand, J., Effects of differential shrinkage, creep, and properties of the contact surface on the strength of composite slabs of old and new concrete, Bulletin No. 147, Department of Structural Mechanics and Engineering, Royal Institute of Technology, Stockholm, 131 pp., 1987 [in Swedish].
30. Schäfer, H.G., Block, K., and Drell, R., Oberflächenrauheit und Haftverbund, *Deutscher Ausschuss für Stahlbeton DafStb*, **456**, 75–94, 1996.
31. Takuwa, I., Shitou, K., Kamihigashi, Y., Nakashima, H., and Yoshida, A., The application of water-jet technology to surface preparation of concrete structures, *Journal of Jet Flow Engineering*, **17**(1), 29–40, 2000.
32. Mainz, J. and Zilch, K., Schubtragfähigkeit von Betonergänzungen an nachträglich aufgerauhten Betonoberflächen bei Sanierungs- und Ertüchtigungsmassnahmen, Research Report, Technical University Munich, Germany, February 1998.
33. Tschegg, E.K., Ingruber, M., Surberg, C.H., and Münger, F., Factors influencing fracture behavior of old-new concrete bonds, *ACI Materials Journal*, **97**(4), 447–453, 2000.
34. Gulyas, R.J., Wirthlin, G.J., and Champa, J.T., Evaluation of keyway grout test methods for precast concrete bridges, *PCI Journal*, **40**(1), 44–57, January–February 1995.
35. Block, K. and Porth, M., Spritzbeton auf carbonatisiertem Beton – Haftzugfestigkeit bei nachträglichem Aufspritzen, *Beton 7*, 299–302, 1989.
36. Delatte, N.J., Wade, D.M., and Fowler, D.W., Laboratory and field testing of concrete bond development for expedited bonded concrete overlays, *ACI Materials Journal*, **97-M33**, 272–280, May–June 2000.
37. Vaysburd, A.M. and McDonald, J.E., An evaluation of equipment and procedures for tensile bond testing of concrete repairs. Technical Report REMR-CS-61, US Army Corps of Engineers, Waterways Experiment Station, Vicksburg, Mississippi, USA, 75 pp., 1999.
38. Zhu, Y., Evaluation of bond strength between new and old concrete by means of fracture mechanics method, Bulletin No. 157, Department of Structural Mechanics and Engineering, Royal Institute of Technology, Stockholm, 102 pp., 1991.
39. Zhu, Y., Effect of surface moisture condition on bond strength between new and old concrete, Bulletin No. 159, Department of Structural Mechanics and Engineering, Royal Institute of Technology, Stockholm, 27 pp., 1992.
40. Li, S.E., Geissert, D.G., Frantz, G.C., and Stephens, J.E., Freeze-thaw bond durability of rapid-setting concrete repair materials, *ACI Materials Journal*, **96-M31**, 241–249, March–April 1999.
41. Saucier, F. and Pigeon, M., Durability of new-to-old concrete bondings. In *Proceedings of ACI International Conference on Evaluation and Rehabilitation of Concrete Structures and Innovations in Design*, Hong Kong, December (ACI SP-128), pp. 689–705, 1991.
42. ACI Committee 546, Guide for Repair of Concrete Bridge Superstructures (ACI 546.1 R-80), American Concrete Institute, Detroit, 20 pp., 1980.
43. Silfwerbrand, J., The influence of traffic-induced vibrations on the bond between old and new concrete, Bulletin No. 158, Department of Structural Mechanics and Engineering, Royal Institute of Technology, Stockholm, 78 pp., 1992.

44. FIP Federation Internationale de la Précontrainte, Shear at the Interface of Precast and In-Situ Concrete, Technical Report, August 1978.
45. de Souza, R.F.F. and da Silva Appleton, J.A., Assessing the structural performance of repaired reinforced concrete members. In *Proceedings ICPCM*, Cairo, Egypt, February 2003.
46. Paulsson, J. and Silfwerbrand, J., Durability of repaired bridge deck overlays – Effects of deicing salt and freeze-thaw cycles, *Concrete International*, **20**(2), 76–82, February 1998.
47. Smithson, L.D. and Whiting, J.E., Rebonding delaminated bridge deck overlays, *Concrete Repair Digest*, 100–101, June–July 1992.
48. Manning, D.G., Effects of traffic-induced vibrations on bridge-deck repairs, NCHRP Synthesis No. 86, Transportation Research Board, Washington DC, 40 pp., 1981.
49. Van Mier, J.G.M., Nooru-Mohammed, M.B., and Timmers, G., Experimental study of shear fracture and aggregate interlock in cement-based composites, *Heron*, **36**(4), 8–30, 1991.
50. Robins, P.J. and Austin, S.A., A unified failure envelope from the evaluation of concrete repair bond test, *Magazine of Concrete Research*, **47**(170), 57–68, March 1995.
51. Weber, M., Mechanischer Verbund zwischen Beton verschiedenen Alters mittels Kunststoffen, Dissertation, Fakultät für Bauingenieur- und Vermessungswesen der Universität Karlsruhe, Karlsruhe, 1971.
52. El-Rakib, T.M., Farahat, A.M., El-Degwy, W.M., and Shaheen, H.H., Shear transfer parameters at the interface between old and new concrete. In *Proceedings ICPCM*, Cairo, Egypt, February 2003.
53. Chen, P., Fu, X., and Chung, D.D.L., Improving the bonding between old and new concrete by adding carbon fibres to the new concrete, *Cement and Concrete Research*, **25**(3), 491–496, 1995.
54. Emberson, N.K. and Mays, G.C., Significance of property mismatch in the patch repair of structural concrete. Part 1: Properties of repair systems, *Magazine of Concrete Research*, **42**(152), 147–160, September 1990.
55. Austin, S., Robins, P., and Pan, Y., Shear bond testing of concrete repairs, *Cement and Concrete Research*, **29**, 1067–1076, 1999.
56. Lacombe, P., Beaupré, D., and Pouliot, N., Rheology and bonding characteristics of self-levelling concrete as a repair material, *Materials and Structures*, **32**, 593–600, October 1999.
57. Sato, R., Recent technology of concrete pavement in Japan. In *Proceedings, S. Nagataki Symposium on Vision of Concrete: 21st Century*, Part of the 4th CANMET/ACI/JCI International Symposium on Advances in Concrete Technology, Tokushima, Japan, Vol. 1998.6, pp. 71–85, 1998.
58. Paulsson, J., Effects of repairs on the remaining life of concrete bridge decks, Bulletin No. 27 (Licentiate Thesis), Department of Structural Engineering, Royal Institute of Technology, Stockholm, Sweden, 238 pp., 1997.
59. Paulsson-Tralla, J., Service life of repaired concrete bridge decks, Bulletin No. 50 (Ph.D. Thesis), Department of Structural Engineering, Royal Institute of Technology, Stockholm, Sweden, 244 pp., 1999.
60. Swedish National Board of Housing, Building and Planning and AB Svensk Byggtjänst, *BBK 94. Concrete Structures, Part 1 – Structural Design, Handbook.* Karlskrona and Stockholm, Sweden, 185 pp., 1994 [in Swedish].
61. American Concrete Institute, ACI 318, Building Code Requirements for Structural Concrete (ACI 318-99) and Commentary (ACI 318R-99). Farmington Hills, Michigan, USA, 391 pp., 1999.
62. Comité Euro-International du Béton & Federation Internationale de la Precontrainte, *CEB-FIP Model Code 1990*. Thomas Telford Services, London, 437 pp., 1993.
63. European Committee for Standardization, Eurocode 2. Design of Concrete Structures. 2nd draft. Brussels, Belgium, January 2001.
64. CEN (Comité Européen de Normalisation), prEN 1504-3, Products and Systems for the Protection and Repair of Concrete Structures – Definitions, Requirements, Quality Control and Evalution of Conformity – Part 3: Structural and Non-Structural Repair, Brussels, Belgium, March 2001.
65. Swedish National Road Administration, General Technical Regulations for Bridges, Publication No. 2004:56, Borlänge, Sweden, 2004 [in Swedish].

Further Reading

ACI Committee 555, Removal and reuse of hardened concrete (ACI 555 R-01), *ACI Materials Journal*, **99**(3), 300–325, May–June 2002,

Bernard, O., Comportement à long terme des elements de structure formés de bétons d'âges différents. Thesis No. 2283, Department of Civil Engineering, EPFL, Lausanne, Switzerland, 189 pp., 2000.

British Standards, BS 6319, Part 4, Slant Shear Test Method for Evaluating Bond Strength of Epoxy Systems, British Standard Institution BSI, London, 1984.

Chausson, H., Durabilité des rechargements minces en béton: Relation entre leur décollement, leur fissuration et leur renforcement par des fibres. Ph.D. Thesis, Université Paul Sabatier, Toulouse, France, 198 pp., 1997 [in French].

Fédération Internationale de la Précontrainte, Shear at the interface of precast and in-situ concrete. Guide to good practice, Wexham Springs, Slough, 31 pp., 1982.

Ingvarsson, H., and Eriksson, B., Hydrodemolition for bridge repairs, *Nordisk Betong (Stockholm)*, **2–3**, 49–54, 1988.

Jonasson, J.-E., Computer programs for non-linear analyses of concrete in the view of shrinkage, creep and temperature. Research Report No. 7:77, Swedish Cement and Concrete Research Institute, Stockholm, Sweden, 161 pp., 1977 [in Swedish].

Julio, E.N.B.S., Branco, F.A.B. Silva, V.D., Concrete-to-concrete bond strength. Influence of the roughness of the substrate surface. *Construction and Building Materials*, **18**, 675–681, 2004.

Silfwerbrand, J., Theoretical and experimental study of strength and behaviour of repaired concrete bridge decks summary, Bulletin No. 149, Department of Structural Mechanics and Engineering, Royal Institute of Technology, Stockholm, Sweden, 16 pp., 1987.

Silfwerbrand, J., Concrete repair with shotcrete. Tests on beams under static and fatigue load, Bulletin No. 153, Department of Structural Mechanics and Engineering, Royal Institute of Technology, Stockholm, 77 pp., 1988 [in Swedish].

Silfwerbrand, J. (1989), Concrete repair with shotcrete. In *Proceedings, IABSE Symposium on Durability of Structures*, Lisbon, September, pp. 785–790, 1989.

Silfwerbrand, J., Concrete overlays, Report No. 10, 3rd Edition 10, Chair of Structural Mechanics and Engineering, Department of Structural Engineering, Royal Institute of Technology, Stockholm, Sweden, 65 pp., 1997 [in Swedish].

Silfwerbrand, J., Stresses and strains in composite concrete beams subjected to differential shrinkage, *ACI Structural Journal*, **94**(4), 347–353, 1997.

Silfwerbrand, J. and Sundquist, H., Drift, underhåll och reparation av konstbyggnader. Rapport nr 53, Brobyggnad, institutionen för byggkonstruktion, KTH, Stockholm, 262 pp., 1999.

Strömdahl, C., The history of hydrodemolition, *Concrete Engineering International*, **4**(8), 32–36, 2000.

Chapter 5
Structural Behaviour

E. Denarié, J. Silfwerbrand and H. Beushausen

Abstract Bonded cement-based material overlays and their substrates constitute a *hybrid or composite structural system*. The interaction of these two material layers (with different ages), with each other, with the external boundary conditions (foundations, supports) and possible joints, and under loading, defines the *structural behaviour* of this composite system.

The main *actions* governing this structural behaviour are (1) the differential deformations of the two layers due to autogenous shrinkage, thermo-mechanical effects due to cement hydration and/or external climatic influences and drying shrinkage, (2) settlements, and/or (3) imposed forces (dead loads, traffic loads). These actions give rise to stresses which depend on the stiffness of the substrate with respect to the new layer, and that eventually may result in failure either by transverse crack propagation, by debonding or both.

Composite structures formed of building materials of different kinds and ages are very common: slabs on grade, steel-concrete, wood-concrete, concrete-concrete, concrete repairs, cement-based overlays, etc. However, until now, the causes of distress in these composite structures and their design were mostly addressed by empirical approaches. This document shows how a common approach might treat all these applications on the unique basis of the mechanical description of the behaviour of composite structural members under restrained shrinkage.

E. Denarié
MCS-IS-ENAC, Ecole Polytechnique Fédérale de Lausanne, Switzerland

J. Silfwerbrand
Swedish Cement and Concrete Research Institute (CBI), SE-100 44 Stockholm, Sweden

H. Beushausen
Department of Civil Engineering, University of Cape Town, Rondebosch, South Africa

5.1 Introduction

Bonded cement-based material overlays and their substrates constitute a *hybrid or composite structural system.* The interaction under loading of these two material layers (with different ages), with each other and with the external boundary conditions (foundations, supports) and possible joints defines the *structural behaviour* of this composite system.

Composite structures formed of building materials of different kinds and ages are very common: slabs on grade, steel-concrete, wood-concrete, concrete-concrete, concrete repairs, cement-based overlays, etc. However, until now, the causes of distress in several of these composite structures and their design were mostly addressed by means of empirical approaches. This document shows how a common approach originally proposed by Silfwerbrand [1, 2] can treat all these applications on the unique basis of the mechanical description of the behaviour of composite structural members under restrained shrinkage.

Composite structural members are subjected to *actions* such as: (1) the differential deformations of the two layers due to autogenous shrinkage, thermo-mechanical effects due to cement hydration and/or external climatic influences and drying shrinkage, (2) settlements, and/or (3) imposed forces (dead loads, traffic loads). These actions give rise to stresses which depend on the stiffness of the substrate with respect to the new layer, and that eventually may result in failure either by transverse crack propagation, by de-bonding or both. The early age effects such as the thermo-mechanical effects are difficult to approximate with simplified approaches, due to the rapidly changing material properties.

There is a *complex interaction* between the structural behaviour and the modes of failure. If de-bonding can be prevented, the hybrid structure behaves in a *monolithic* way and structural failure will happen by transverse cracking. Otherwise, if de-bonding occurs, near cracks or joints, complex failure patterns can be obtained.

In the case of restrained deformations of the overlay, a simplified approach is to summarize the parameters related to the structure using the notion of *degree of restraint* of the overlay. Maximum restraint is obtained when all degrees of freedom (flexural and axial) of the overlay are blocked, i.e. the worst case scenario. The degree of restraint represents the *loading level* of the overlay with respect to maximum restraint. This approach has been investigated by several authors with the hypothesis of elastic materials. In the most general case, the structural response of a hybrid element can only be determined by a finite element simulation taking into consideration all relevant boundary conditions and material properties. The *viscoelastic behaviour* of the overlay, and if applicable, of the substrate, significantly contributes to the stress relaxation under imposed deformations. It can be taken into consideration by comprehensive numerical analyses, or by means of simplified analytical models.

Both the simplified approach based on the degree of restraint and the advanced numerical simulations can be used to *predict* the structural performance of composite systems, under imposed deformations of the overlay, with the *aim to select appropriate combinations of materials, thicknesses, reinforcements and joints.*

5.2 Actions

Note: The word "Actions" has to be understood here in a broad sense encompassing imposed forces and dead loads, as well as imposed displacements or deformations.

At early age and during their service life, overlays and composite members are subjected to various kinds of actions such as:

- *Imposed deformations* – autogenous effects at early age, drying shrinkage, thermo-mechanical effects associated with hydration of binders (temperature gradients) at early age or to temperature gradients over the long term (see Chapter 3).
- *Imposed displacements* – settlements, seismic loads.
- *Imposed forces* – braking forces of vehicles, moving (fatigue) or gravity loads – permanent loads in storage facilities (see Chapter 7).

The aim of this chapter is to provide an understanding of the mechanical behaviour in the case of restrained deformations of the overlay, rather than provide a comprehensive design method. A more detailed description of the deformations of overlay materials is given in Chapter 3. The overall actions to be considered for design as well as usual design methods are discussed in Chapter 7, for which this chapter can be seen as background.

5.3 Performance of Composite Structures

Composite structural components are formed of layers of different properties and age connected in a more or less monolithic way. The following applications belong to this category:

- steel-concrete composite constructions;
- partial replacement of existing concrete superstructures (kerbs on bridges);
- overlays on old concrete substrates (beams, slabs, columns);
- overlays on precast slabs; and
- pavements.

In the case of overlays, the overall performance of a composite system after the casting of a new layer on an existing substrate must be evaluated in the following terms:

- the protective function of the new layer and its serviceability; and
- the load-bearing capacity of the hybrid component and its behaviour at serviceability and ultimate (limit) state.

Figure 5.1, after Bernard [3], shows the evolution of the performance of a typical hybrid structural component formed of a new layer cast on an existing substrate. At early age, the partial restraint of the deformations of the new layer gives rise to

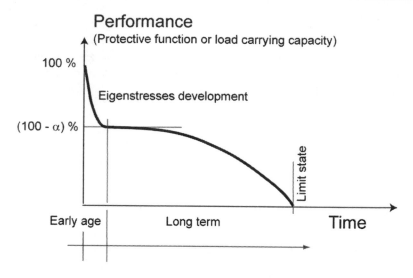

Fig. 5.1 Performance of composite structures [3]

eigenstresses that decrease with time due to the viscoelastic behaviour of the materials. The value of these eigenstresses can be considered as a "penalty" imposed on the intrinsic performance of the new layer, in terms of a reduction of the tensile stress that can be supported by this layer under service loads. This drop of the initial performance of the new layer must be taken into consideration in the calculation of the mechanical behaviour of the composite system. The mechanical behaviour of a composite structural member under loads can be calculated by means of simple well known theories from continuum mechanics and strength of materials. However the resistance to be considered must take into consideration the history of the various layers at early age or due to shrinkage by means of eigenstresses. Finally, eigenstresses tend to vanish with time as a consequence of the decreasing shrinkage rate of the overlay and viscoelastic response of the materials. The time to decrease the maximum eigenstresses by a factor 2 is however counted in years rather than in month.

5.4 Different Forms of Restraint and Effect of Joints

Restraint can be defined as all mechanical effects that counteract the deformations of a given body.

- At the micro level of heterogeneous materials such as concretes, the aggregates act as a restraint to the deformations of the cement paste under all kinds of shrinkage. The effect of this "internal restraint" on the onset of stresses and microcracks was numerically demonstrated among others by Sadouki [4]. The same applies to Steel Fibre Reinforced Concretes, for which the fibres may significantly decrease the apparent shrinkage.

- The frictional forces on the subgrade of slabs on grade are another major source of restraint extensively studied in the literature. This effect is commonly described by interface models with adapted Mohr–Coulomb models.
- The flexural and axial stiffness of the substrate on which an overlay is applied act as a restraint. The combined axial and flexural stiffness of the composite system formed of substrate and overlay have to be considered to determine the degree of restraint of the overlay.
- The restraint by dowels is the main effect governing the structural behaviour of steel-concrete composite beams, as well as any other combination of different layers of different materials, linked by connectors.
- Reinforcement bars and in some cases formworks also act as a restraint to the deformations of cementitious materials at early age and long term, according to Bernard [3].
- Finally, the static system of a structural member also provides a restraint which acts on different degrees of freedom.

On the other hand, joints (artificial or cracks) in a composite structure act as a local release of the restraint, with varying consequences on the overall behaviour, positive or negative. The combination of the various kinds of restraint and of the joints defines the *kinematic system*.

5.5 Mechanical Behaviour of Composite Structures with Cementitious Materials of Different Ages

5.5.1 Overview of Existing Analytical Models

In what follows, the focus will be put on the description of the mechanical response of composite structural members under differential shrinkage. The existing analytical models can be classified on the basis of three main features:

- consideration of the flexural degree of freedom (effect of curvature);
- consideration of partial debonding;
- consideration of the viscoelastic behaviour of overlay and substrate.

There are two major internationally accepted analytical theories for the modelling of bonded concrete overlays. These are the theories presented by Birkeland [5], which does not consider debonding, and the one from Silfwerbrand [1, 6] which does. The models presented by Alonso-Junghanns [7], Silfwerbrand [1, 2], Bernard [3], Denarié et al. [8], and Hartl [9, 10] are essentially adaptations and modifications of Birkeland's theory. The prestress analogy (introduction of concentrated forces at the free ends of a composite member to model the effect of shrinkage) which is applied in the models based on Birkeland's theory, is convenient to use, especially because it is easy to understand and easy to apply.

Beushausen [11] questioned both: (1) existing analytical models based on the prestress analogy, in their way to describe the introduction of forces in a composite member subjected to differential shrinkage of an overlay, and (2) the use of the Bernoulli principle in the derivation of most models. This author presented a semi-empirical analytical model for the analysis of restrained overlay shrinkage stresses based on localised, i.e. non-linear strain and stress conditions inside the composite member. Extensive strain measurements were realized on composite specimens with various geometries and aspect ratios (curvature free or prevented), with different surface preparations of the substrate. As expected, for deep composite elements with aspect ratio $l/h < 1$, Bernoulli's principle of plane sections remaining plane after being stressed does not apply [12–14].

On a general basis, one must keep in mind that shrinkage of an overlay corresponds to an imposed displacement, not an imposed force or imposed distributed forces. It is clear that the displacement due to shrinkage is partly restrained on all the length of the interface with the substrate. This however does not imply that the resulting reaction of the substrate leads to constant shear forces along the interface. At the contrary, for "slender geometries" with l/h superior to 5, following what was shown by Jonasson [15] and Haardt [16], a simple finite element simulation of a drying overlay on a substrate leads to concentrated shear forces in the interface, at the extremities of the member, similar to the predictions of the "prestress analogy", which turns out to be well adapted to simply represent a composite member with shrinkage of the overlay.

In what follows, the two main approaches will be referred to as:

- *Perfect bond* by Silfwerbrand [1, 2], further generalized by Bernard [3] and modified by Denarié et al. [8].
- *Partial debonding*, from Silfwerbrand [1, 17].

5.5.2 Normal Stresses Due to Differential Shrinkage in Composite Beams with Complete Bond

Stresses due to restrained movements can principally be computed by the product of three factors according to the following equation:

$$\text{Stress} = \text{stiffness} \times \text{free strain} \times \text{degree of restraint}$$

Consequently, all three factors are equally important. The stiffness is dependent on modulus of elasticity but also on creep or relaxation. The free strain is the strain that a completely free member would develop due to thermal or moisture changes, shrinkage, or any other internal or external source causing volumetric change of the member material. The degree of restraint μ defines the conditions of restraint as the ratio between the actual stress σ_{rest} taking into consideration the effective stiffness

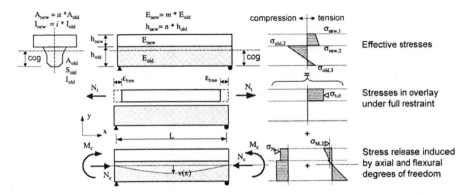

Fig. 5.2 Eigenstresses in a statically determinate composite member, combination of axial (σ_N) and flexural ($\sigma_{M,2}$) release effects (adapted from [3])

of the composite structure and the stress σ_{full} that would occur in a totally restrained composite structure:

$$\mu = \frac{\sigma_{rest}}{\sigma_{full}} \tag{5.1}$$

Restraints can be associated to all degrees of freedom of a structure. For a composite beam, two degrees of freedom can be realized, one axial, one flexural. *Many structural engineers are not aware of the fact that complete bond between overlay and substrate does not necessarily cause complete restraint in the repaired concrete structure.* The reason is that the stiffness of the remaining part of the old structure is not infinite. The striving of the overlay to contract is only partly prevented by the remaining part of the old structure. The absence of a complete restraint leads to substantial stress reductions. Combined with creep these reductions will limit the maximum tensile stress below the tensile strength and, hence, explain the absence of shrinkage cracking. The easiest way to explain this is by studying a composite beam exposed to differential shrinkage. This analysis was done first by Silfwerbrand [1] for rectangular layers of equal width, and generalized by Bernard [3] to distinguish the contribution of the various degrees of freedom, for arbitrary cross sections of the two layers, as shown in Figure 5.2, for a statically determinate beam, with $\sigma_{new,2}$ [MPa]: tensile stress in the new layer at the interface, μ: degree of restraint, ε_{free}: mean shrinkage strain in the new layer, cog: centre of gravity of the composite section, h_{new}, cog_{new}, A_{new}, S_{new}, I_{new}, resp. height, centre of gravity, area, static moment and inertia of the new section (overlay), h_{old}, cog_{old}, A_{old}, S_{old}, I_{old}, resp. height, centre of gravity, area, static moment and inertia of the old section (substrate), and $n = h_{new}/h_{old}$, $m = E_{new}/E_{old}$, $a = A_{new}/A_{old}$, $i = I_{new}/I_{old}$, I_{comp} = inertia of the composite section.

The degree of restraint is calculated under the following hypotheses: linear-elastic material behaviour, Poisson's ratio $\nu = 0$, cross-section of the new layer is a rectangle, the cross-section of the substrate can be of any shape, and plane sections remain plain (hypothesis of Bernoulli), perfect bond between new layer and substrate. *As a consequence, the calculations only apply for the case of slender composite beams ($l/h < 5$). For deep elements such as walls on slabs, with $l/h < 5$, similar principles can be applied. The calculation of the strains and stresses has however to be adapted.*

The principle of the analysis (prestress analogy) consists in determining the tensile force N_t that is necessary to compensate the free shrinkage deformation ε_{free} in the new layer. The tensile force is balanced in the composite member by a compressive force N_c and a bending moment M_c acting at the centre of gravity (cog) of the composite section. The stress state in the composite element is determined by the superposition of the resulting effects of N_t, N_c and M_c on the composite cross section. The resulting stress is the sum of the stress in the case of a total restraint, corresponding to the effort N_t, plus the relaxing effects due to the axial (N_c – stress σ_N) and flexural (M_c – stress $\sigma_{M,2}$) degrees of freedom. The axial release effect inducing the stress σ_N is constant throughout the overlay thickness. This is not the case of the flexural release effect inducing the stress σ_M which is maximum at the top of the overlay and has its smallest value in the overlay nearby the interface. *In order to estimate the maximum overall stress in the overlay, the flexural release effect is evaluated at its location of minimal value, i.e., nearby the interface, and is noted $\sigma_{M,2}$.*

In his original analysis, Bernard [3] proposed to define the degree of restraint as the sum of "1" plus the axial factor μ_N plus the flexural factor μ_M: both factors are negative in their individual expressions.

In the following, this approach is slightly modified, to define the degree of restraint as "1" minus the axial release minus the flexural release (Equation (5.2)), to associate the notion of degree of restraint to the effect of release of the stresses for each degree of freedom available, in a more straightforward way [8].

$$\mu = \frac{\sigma_{new,2}}{\sigma_{full}} = \frac{\sigma_{full} + \sigma_N + \sigma_{M,2}}{\sigma_{full}} = 1 - \mu_N - \mu_M \tag{5.2}$$

The individual expressions of the release factors are adapted from [3]:

$$\mu_N = \frac{-\sigma_N}{\sigma_{full}} = \frac{ma}{ma+1} = \frac{1}{1+1/ma} = \frac{1}{1 + \frac{E_{old} \cdot A_{old}}{E_{new} \cdot A_{new}}} \tag{5.3}$$

$$\mu_M = \frac{-\sigma_{M,2}}{\sigma_{full}} = \frac{N_t(cog_{new} - cog)}{W_2} \frac{1}{E_{new} \cdot \varepsilon_{free}}$$

$$= \frac{A_{new} \cdot (cog_{new} - cog) \cdot [m \cdot (h_{old} - cog)]}{[I_{old} + A_{old}(cog - cog_{old})^2 + m \cdot (I_{new} + A_{new}(cog_{new} - cog)^2)]} \tag{5.4}$$

With W_2: resisting moment of the composite section, at the level of the interface, I_{comp}: inertia of the composite section, y_2, lever arm between cog and interface (location of stress $\sigma_{new,2}$)

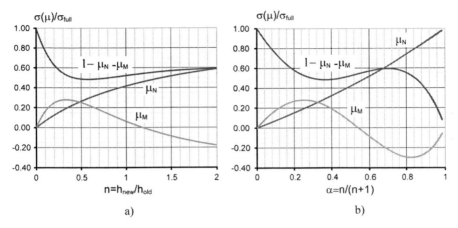

Fig. 5.3 Axial and flexural effects on restraint as a function of (a) $n = h_{new}/h_{old}$ and (b) $\alpha = h_{new}/(h_{old} + h_{new})$, with $m = 0.71$ and $h_{old} = h_{new}$, after [8]

$$cog = \frac{S_{old} + m \cdot S_{new}}{A_{old} + m \cdot A_{new}} \tag{5.5}$$

$$W_2 = \frac{I_{comp}}{y_2}$$

$$= \frac{[I_{old} + A_{old}}{(cog - cog_{old})^2 + m \cdot (I_{new} + A_{new}(cog_{new} - cog)^2)]} m \cdot (h_{old} - cog) \tag{5.6}$$

It is worth mentioning that the release factor μ_N associated to the axial degree of freedom corresponds to the approach proposed by Alonso Junghans [7].

The graphical representation of equations (5.2) to (5.6) is shown in Figure 5.3 for $m = 0.71$ ($E_{new} = 25$ GPa and $E_{old} = 35$ GPa) and rectangular sections of similar width for the old and new layers ($h_{old} = h_{new}$). The axial release μ_N increases in a monotonic way when the thickness of the new layer increases. The flexural release μ_M first increases, reaches a maximum and then decreases down to 0 when the centre of gravity (COG) of the composite cross section enters the new layer, to become then negative. Two representations are shown: (a) with the ratio of the layer thicknesses n, and (b) with the ratio α between the thickness of the new layer and the total thickness of the composite section as x-axis.

The range of most cases encountered in practice corresponds to the domain shown in Figure 5.3a with parameter n on the x-axis. For the chosen set of parameters, the global restraint varies in a significant way for values of n smaller than 0.3. For values of n larger than 0.3, the overall degree of restraint is almost constant and equal to 0.5 to 0.6.

Silfwerbrand [1, 2] generalized this method for various combinations of materials, as shown in Figure 5.4a, and to different boundary conditions [17]. When the

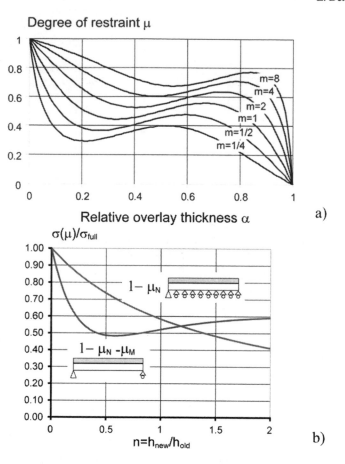

Fig. 5.4 (a) Degree of restraint as a function of relative thickness and relative stiffness, after [1, 2]. (b) Effect of statical boundary conditions with $m = 0.71$ and $h_{old} = h_{new}$

flexural degree of freedom is almost or fully blocked (for example in the middle spans of hyper static multiple span beams), only axial release acts and the degree of restraint is given by $\mu = 1 - \mu_N$, as illustrated in Figure 5.4b.

Concrete viscoelasticity has a beneficial effect on the developed stresses due to differential shrinkage, in terms of relaxation. An engineering approach to estimate this effect is to replace the modulus of elasticity E_{new} of the overlay with a fictitious modulus of elasticity $E^*_{new} = E_{new}/(1+\phi_{new})$ and E_{old} with $E^*_{old} = E_{old}/(1+\phi_{old})$, where ϕ_{new} and ϕ_{old} are the creep coefficients of the overlay and base concrete, respectively. The maximum tensile stress in the overlay may be computed by the following equation according to Silfwerbrand [2, 18]:

$$\sigma_{max} = \frac{m^*(1-\alpha)(m^*(1-\alpha)^3 + \alpha^2(3+\alpha))}{m^* + (m^*-1)(m^*(1-\alpha)^4 - \alpha^4)} \cdot \frac{E_{new}}{1+\varphi_{new}} \cdot \varepsilon_{free} = \mu \cdot \frac{E_{new}}{1+\varphi_{new}} \cdot \varepsilon_{free}$$

$$(5.7)$$

with

$$m^* = \frac{E_{old}^*}{E_{new}^*} = \frac{E_{old}/(1+\varphi_{old})}{E_{new}/(1+\varphi_{new})} = \frac{E_{old}}{E_{new}} \cdot \frac{1+\varphi_{new}}{1+\varphi_{old}} \approx 1 + \varphi_{new}$$

since $E_{new} \approx E_{old}$ and $\phi_{old} \approx 0$.

Since the development of drying shrinkage starts as soon as the curing of the new-cast overlay is ended, also the stresses due to differential shrinkage start to develop at an early stage. This means that also the creep starts at an early stage, before the concrete is mature. This will in turn lead to a large amount of creep. Creep coefficients of 5 to 6 have been measured [2, 18]. The beneficial influence of creep is shown in the following example of computation of stresses due to differential shrinkage in a repaired concrete beam.

Geometrical and material data: $\alpha = 2/7$ (≈ 0.286), $E_{new} = E_{old} = 35$ GPa, $\varepsilon_{free} = 0.45$ mm/m.

Case (a) Full restraint $\mu = 1 \rightarrow \sigma_{full} = E_{new} \cdot \varepsilon_{free} = 15.8$ MPa.

Case (b) Beam theory with neglected creep $\rightarrow m = 1$, $\mu = 0.452$ and $\sigma_{full} = 7.12$ MPa.

Case (c) Beam theory with creep coefficients $\phi_1 = 4$ and $\phi_2 = 0 \rightarrow m = 0.2$, $\mu = 0.733$, and $\sigma_{full} = 2.31$ MPa.

The maximum normal stress diminishes from 16 to 2.3 MPa. The latter value is of the same magnitude as the tensile strength of concrete. It explains why some overlays are crack-free while cracks are visible in others. The creep factors assumed in this example are theoretical estimates to illustrate the possible effect of relaxation of stresses due to viscoelastic effects.

One must however also mention that the resistance of cementitious materials subjected to long term tensile actions with low loading rates such as shrinkage is significantly lower than the resistance obtained from quasi static tensile tests. This effect has to be considered when evaluating the risk of cracking at early age or long term in bonded overlays.

Beushausen et al. [19] estimated the relaxation in overlays from experimental tests on composite concrete/concrete members. Strain measurements on free and restrained materials and time of cracking were used to determine the relaxation at a given time. Values of 40 to 50% were found. More generally speaking, these authors propose an incremental method to determine the relaxation factor as a function of time.

A more precise analytical approach of the effect of linear viscoleasticity on the relaxation of stresses in composite members can be found in the multilayer theory from Huet [20], adapted from the elastic case from Siestrunck et al. [21].

Finally, the viscoelastic response depends on the stress level above certain limits. There is no general agreement in the literature on the threshold value above which viscoelastic response would be significantly influenced by the stress level. Gustch et al. [22] and Horimoto et al. [23] tend to show that the relaxation under tension is not significantly influenced by the stress level. Denarié et al. [24] showed that relaxation starts to deviate from a linear response at load levels around 50% in Wedge Splitting Tests. However, for this specific testing geometry with a strong strain gradient, the material enters the softening domain at the notch tip for load levels close to 50%. At this level, the deviation from non-linearity corresponds to the beginning of activation of the tensile strain softening (equivalent load level 100% under uniaxial tension).

Under creep, it is well known that deviation from a linear viscoelastic response start to be clear above 40 to 50% under compression and tension.

On the other hand, for high degrees of restraint, stresses can very quickly be in the range of the quasi-static tensile strength. At those levels of loading, linear viscoelastic models do not apply anymore and any substantial increase in the viscoelastic response is likely to be due to ongoing damage.

Reliable relaxation tests under tension, at various load levels pre peak are extremely difficult to perform and constitute an interesting and necessary challenge for future research.

5.5.3 Shear Stresses Due to Differential Shrinkage in Composite Beams

The simple beam model (described in Section 5.5.2) assuming complete bond between overlay and base (substrate) is easy to use and gives decent estimations of normal stresses in the interior parts of overlay and base. However, in reality horizontal cracking leading to debonding sometimes is observed especially in vicinity to the vertical borders and joints of the repaired structure. In order to estimate the probability for debonding or explain its presence if it already exists, we need to be able to estimate shear stresses along the interface between overlay and substrate. Jonasson [15] has developed a simple model that also has been used by FIP [25]. Jonasson made computer investigations and found that the shear stress has its maximum magnitude at the boundary of the structure and diminishes approximately linearly to zero at a distance approximately equal to three times the thickness of the overlay (Figure 5.5).

By utilizing the symbols presented above, the maximum shear stress τ_{max} is given by the following equilibrium equation:

$$\tau_{max} = \frac{2}{3} \cdot \frac{F}{b\alpha h} \tag{5.8}$$

where αh and b are the thickness and width of the overlay, respectively, and F is the resultant to the normal stresses in the overlay, i.e.

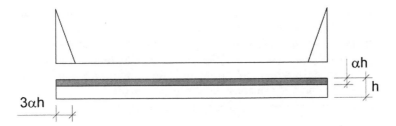

Fig. 5.5 Shear stress distribution according to [15]

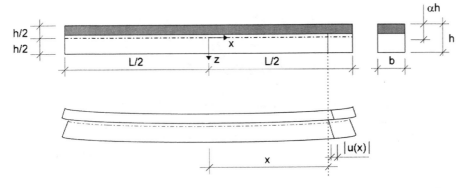

Fig. 5.6 Composite beam with incomplete bond between overlay and base (substrate) [1, 6]

$$F = \int_{\alpha h} \sigma(z) \cdot b \cdot dz \qquad (5.9)$$

For the simply supported or free composite beam (Figure 5.3), the maximum shear stress could be computed by the following equation (if creep is considered):

$$\tau_{max} = \frac{2}{3} \cdot \frac{m(1 - 4\alpha + 6\alpha^2 - 3\alpha^3 + (m-1)(1-\alpha)^4)}{m + (m-1)(m(1-\alpha)^4 - \alpha^4)} \cdot \frac{E_1}{1 + \phi_1} \varepsilon_{cs} \qquad (5.10)$$

5.5.4 Normal and Shear Stresses Due to Differential Shrinkage in Composite Beams with Incomplete Bond

A shortcoming with the simple beam model described in Section 5.3 is that it leads to constant normal stresses along the beam in cases with symmetrical boundary conditions. Since the shear stress is the derivative of the normal stress with respect to the length coordinate, constant normal stresses result in zero shear stresses. In reality, shear stresses appear in the vicinity to vertical boundaries, vertical joints, and vertical cracks, if any.

Silfwerbrand [1, 6] has developed an engineering beam model considering incomplete bond between overlay and base. By using this model, it is possible not only to compute normal stresses and strains along the beam, but also shear stresses $\tau(x)$ and horizontal slip $u(x)$ along the interface. This model uses a linear relationship between shear stress and horizontal slip between overlay and base, i.e., $\tau(x) = K \cdot u(x)$, where K is a constant of proportionality.

The shear stresses can be computed by the following equation [6]:

$$\tau(x) = -\frac{m\alpha(1-\alpha)(1 - 3\alpha(1-\alpha) + (m-1)(1-\alpha)^3)}{m + (m-1)(m(1-\alpha)^4 - \alpha^4)}$$

$$\cdot \frac{E_1}{1+\phi_1} \cdot \varepsilon_{cs} \cdot \frac{h}{L} \cdot \lambda L \cdot \frac{\sinh \lambda x}{\cosh \lambda L/2} \tag{5.11}$$

$$(\lambda L)^2 = -\frac{m + (m-1)(m(1-\alpha)^4 - \alpha^4)}{m\alpha(1-\alpha)(1 - 3\alpha(1-\alpha) + (m-1)(1-\alpha)^3)} \cdot \frac{1+\phi_1}{E_1} \cdot \frac{KL^2}{h} \tag{5.12}$$

The engineering problem is to estimate the coefficient K and, hence, the governing non-dimensional product λL. Direct tests are difficult since the horizontal slip u is difficult to measure and likely to be dependent on loading rate. The interesting rate is given by the shrinkage velocity and such slow tests are if not practically impossible at least economically impossible. However, the effect of varying values of the product λL could easily be studied (Figure 5.7), with $z^* = z$ coordinate at interface.

Shear stresses increase with increasing λL value and normal stresses approach the constant distribution given by the simple beam model with complete bond (Section 5.3).

Normal stresses can be seen as the reflection of the strains and vice versa. The normal stress vanishes at the beam edges and has its maximum value in the interior parts of the beam. The strain (that contrary to stress is easy to measure) has, consequently, its maximum variation at the edges. This relationship has been verified by strain measurements in two 6 and 8 m long repaired concrete beams subjected to 14 months of differential shrinkage [1]. By considering creep effects and comparing the measuring results with computed strain curves for varying λL values, it was concluded that $\lambda L = 54$ gave the best fit to the 8 m long beam (Figure 5.8). This value has in turn been used to estimate the shear stresses in this beam (Figure 5.9).

The maximum shear stress is estimated to 1.6 MPa, i.e., probably below the shear strength of concrete. This computation result is an explanation to the fact that no cracks were observed in the test beams [1].

The computation result also shows that the simple shear stress computation model [15] described in Section 5.3 gives a fairly good estimation of both maximum shear value and its corresponding triangular shape.

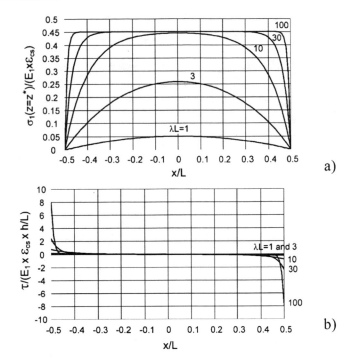

Fig. 5.7 (a) Maximum relative normal stresses ($= \psi$) and (b) relative shear stresses along a simply supported composite beam due to differential shrinkage. $m = 1$, $\alpha = 2/7$, $\phi_1 = \phi_2 = 0$, after [1, 6]

5.6 Experimental Tests

Various authors have performed tests on composite structural members. However, very few have produced results that support or discard models.

5.6.1 Swedish Tests on Mechanically Loaded Concrete Beams

Silfwerbrand [26] carried out tests on simply supported composite concrete beams subjected to a static load at mid span (Figure 5.10). The substrate concrete was sawn from a 70 year old bridge (Skurubron). It had a compressive strength of 85 MPa (measured on drilled cores). The substrate concrete was chipped with pneumatic hammers prior to overlay casting. The measured compressive cube strength of the overlay was 60 MPa. For comparison, additional test beams made of new-cast concrete were included in the investigation. Bonding agents were used in two cases. The beams were tested upside down producing tension in the overlay. As shown in Table 5.1, the composite beams with hammered interfaces developed equally high

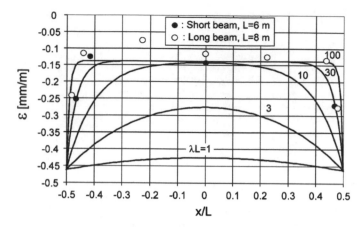

Fig. 5.8 Measured strain values and computed ones, from [6]

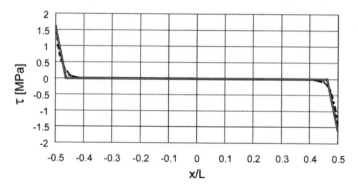

Fig. 5.9 Computed shear stresses for an example based on geometrical and material data for the 8 m long beam [6, 18]. Solid line = simple model proposed by Jonasson [15], dotted line = model for incomplete bond [6]

failure loads and obtained same failure mode as the homogeneous beam. Obviously, the bond between substrate and overlay was sufficient. On the contrary, a smooth, steel grinded interface does not provide sufficient bond to promote composite action. In both beams with this interface, a premature interface failure occurred. After the interface failure, the (reinforced) bottom part of the beam was able to carry an increasing load, but not to the same level as the other beams.

Table 5.1 Tests on mechanically loaded concrete beams [26]

Interface			Static tests	
Preparation	Bonding agent	Other treatment	Failure load (kN)	Failure mode
Pneumatic hammer	No		130	Shear
Pneumatic hammer	Epoxy adhesive		150	Shear
Pneumatic hammer	No	Vacuum treated overlay	130	Shear
None (homogeneous)	No		130	Shear
Steel grinded	No		40 & 70*	Interface & Shear
Steel grinded	Mortar		17 & 69*	Interface & Shear

* The composite beam first cracked at the interface; subsequently, the load could be increased markedly to shear failure in the bottom (reinforced) part of the beam.

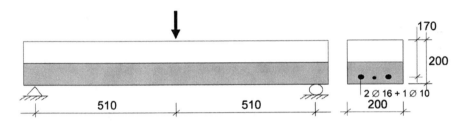

Fig. 5.10 Test specimen and loading case in beam tests (unit: mm) [26]

5.6.2 Swedish Tests on Concrete Beams Subjected to Differential Shrinkage

Differential shrinkage is considered to be an important loading case for concrete overlays. In order to study strain development and cracking, if any, four composite beams of varying length were cast and stored outdoors in a tent for 14 months [1]. The thickness ratio between overlay and total beam height was in all cases 2/7. Two beams were supported on soft airbags providing free curvature whereas the other two were supported on three solid supports preventing curvature (Figure 5.11). Old concrete columns constituted the substrate concrete. Drilled cores obtained a compressive strength of 62 MPa. The substrate concrete was chipped with pneumatic hammers prior to overlay casting. The surface was prewetted during a couple of days but superficially dry at overlay placement. No bonding agents were used. At the end of the tests, the concrete used for the overlays obtained compressive cube strength of 65 MPa. Maximum free shrinkage (measured on 400 × 100 × 100 mm standard prisms) was 0.45 mm/m. Despite the magnitude of the free shrinkage and the bond providing restraint, no cracks were observed, neither in the overlays perpendicular to the beam length nor in the interface. Strain measurements supported

Tests on Composite Beams   Tests on Composite Slabs

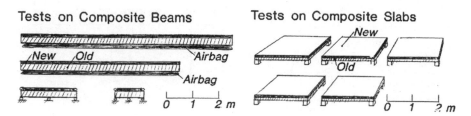

Fig. 5.11 Tests on composite beams [1] and composite slabs [18] subjected to differential shrinkage

composite action. After terminated tests, a bond strength averaging 1.9 MPa was obtained through pull-off tests.

5.6.3 Swedish Tests on Concrete Slabs

In mid 1980s, the water-jet technique was introduced in Sweden. In order to compare it with other methods of concrete removal and substrate preparation, slab tests were carried out at KTH in Stockholm, Sweden [18]. The preparation methods investigated included water-jet technique, pneumatic hammers, and sandblasting. The substrate surface was prewetted during three days but superficially dry at overlay placement. No bonding agent was used. The tests consisted of five composite slabs (Figure 5.12) and two homogeneous ones. The ratio between overlay thickness and total height was 1/3. The substrate concrete was seven months old at time of overlay placement. Differential shrinkage was investigated during six months under indoor conditions. No visible cracking occurred. Displacement measurements supported composite behaviour. At terminated tests, the compressive cube strength exceeded 50 MPa for both substrate and overlay concrete. The free shrinkage (measured on $400 \times 100 \times 100$-mm prisms) of the overlay concrete was 0.6 mm/m. Simultaneously, the shrinkage of the old concrete increased with less than 0.05 mm/m.

In a second phase of the tests (Figure 5.12a), the five composite and the two homogeneous slabs were simply supported at the four corners and loaded to failure with a centrically placed static load.

The base layer was reinforced with 13 Ø 8 mm, $s = 150$ mm in both directions. The obtained load-displacement curves show a similar behaviour for all slabs. All slabs (including homogeneous ones) obtained approximately equally high ultimate loads between 83 and 98 kN (Figure 5.12b). Finally, the bond strength was investigated by pull-off tests (see Section 4.4).

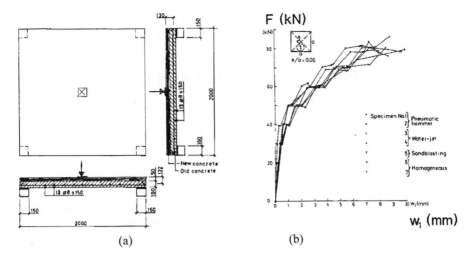

Fig. 5.12 Composite slab loading experiments: (a) experimental layout; (b) load-displacement curves [18]

5.6.4 Tests at EPFL on Composite Beams with Normal Concretes of Different Ages

Bernard [3] tested 14 composite beams with substrate dimensions length/width/thickness of 5400/500/170 mm, hydrojetted with removal of 20 mm of concrete, and new concrete layers between 70 to 170 mm, with various reinforcement ratios, with reinforcement bars or steel fibres. The beams, placed in a controlled environment (20°C, 30% RH) were instrumented with external LVDT's and thermocouples and optical deformation sensors embedded in the substrate and the new layer. The substrates were between 70 and 237 days old when the overlays were cast. The new concretes composition was 300 kg/m^3 CEM I 52.5 R, $W/C = 0.50$ without steel fibres and 350 kg/m^3 CEM I 52.5 R and $W/C = 0.5$, with 80 kg/m^3 steel fibres. The deformational behaviour of the composite members was monitored at early age for 3 to 6 days, and at longer term, up to 270 days, under creep loading. The optical deformation sensors measured total displacements in the central part of the beam (distance 2000 mm). The thermocouples gave the temperature at the same level. The measurements were started when the new concrete was cast and stopped after 100 hours. The main results from this study can be summarized as follows:

- the time lag between the end of chemical swelling and the peak of temperature in the upper part of the new layer and the evolution of strains are well predicted only if the autogenous shrinkage is taken into account in the numerical modelling;
- the autogenous shrinkage is the main phenomenon which induces curvature and tensile stresses at early age in the new layer of tested hybrid elements;

- in the conditions of these tests, the long term residual stresses due to autogenous shrinkage attain approximately 50 % of the maximum tensile stress at early age; and
- the reinforcement in the new layer induces a significant restraint of the overall deformations of the composite member; for high reinforcement ratios of the new layer, this restraint can lead to cracking.

5.6.5 Other Tests

Finally, it is also worth mentioning the works by Chausson [27] that used a steel substrate with a grinded surface to simulate a concrete substrate. This method has significant advantages in terms of reutilization of the substrates for several tests.

5.7 On Restrained Shrinkage Set-ups

Another way to experimentally investigate the effect of restraint on the development of stresses and the risk of cracking, without having to test costly large size structural members, is given by so called "restrained shrinkage testing devices". These devices have been known for many years now, starting with so-called "cracking frames" [28], where the restraint is passive. Active systems were proposed by Kovler [29]. Finally, so called *Temperature-Stress Testing Machines* – TSTM are systems with active restraint by means of closed-loop-testing machines as well as cooling circuits placed around the specimens, to impose temperature conditions at will [30–36].

Such systems can be used either in isothermal conditions to isolate fundamental material properties such as autogenous shrinkage, or to test overlay materials in realistic conditions of restraint and curing, thus providing and "experimental simulation tool".

In a similar way, to induce a high restraint with compact dimensions, Martinola [37] used instrumented ring geometry, inspired from the "Bolomey" ring shrinkage test.

5.8 Numerical Modelling

The numerical modelling of the structural behaviour of bonded concrete overlays and more generally of composite concrete-concrete structural members has been realized in various ways in the past, with either general purpose finite element packages, specifically oriented packages for multi-layer systems, or original numerical solutions of mathematical formulations of the problem. Basically, any finite element package is able to model multi layer systems. The main differences come from the ability of the software to deal with more advanced material properties such as:

- time-dependent evolution of the mechanical properties of the overlay and substrate – maturation;
- comprehensive simulation of the physical processes such as drying, release of heat of hydration, thermal transport;
- viscoelasticity and ageing;
- non-linear crack propagation (smeared crack models or discrete crack models);
- interface between two layers;
- reinforcing bars.

It is impossible to quote all modelling works on hybrid systems however, among the most recent ones, one can mention the following:

Bernard [3, 38, 39] used the finite element package MLS (Multi-Layer Systems) from FEMMASSE [40], with a comprehensive description of thermo-hygro-mechanical processes in multi-layer systems, at early age and long term, including effects of maturation, viscoelasticity, reinforcement in the overlay and non-linear fracture mechanics (smeared crack model) for the bulk of the materials and for interfaces. The significant role of autogenous shrinkage at early age on the structural response of hybrid elements with concretes and the consequences of the restraining effect of reinforcement bars in the new layer were demonstrated. The results include an extensive parametric study on the effect of the overlay thickness and reinforcement ratio on the risk of delamination. Habel [41, 42] used the same package with material models extended to the case of tensile hardening fibre reinforced composites to model composite beams with Ultra-High Performance Fibre Reinforced Concrete (UHPFRC) overlays.

Martinola [37, 43] used FEM package DIANA [44] for thermo-hygro-mechanical calculations with a strong emphasis on moisture transport and self-desiccation, in order to study the risk of cracking and delamination of composite structural systems. Realistic interfacial constitutive laws were used with parameters from experimental tests. Based on comprehensive experimental tests, the hygro-mechanical coupling coefficient was implemented in the code as a function of the moisture content, and the major influence of this variation was demonstrated. A parametric study was performed and recommendations for the design of bonded concrete overlays were given.

Laurence [45] used software CESAR-LCPC to calculate the evolution of stresses at early age and long term in bonded concrete overlays. The modules TEXO, MEXO and HEXO of the finite element package were used to calculate the stresses induced by thermo-chemo-mechanical effects at early age, taking maturation into consideration, for linear elastic materials. In a second step, the effect of drying shrinkage, including water absorption by the substrate, was introduced. The results showed the very significant influence of autogenous shrinkage as well as of the absorption of water by the substrate on the development of stresses in the overlay.

Finite elements computations of composite structures with bonded overlays subjected to drying have been performed by Tran et al. [46] using the FEM code

CAST3M. They take into account the effects of the shrinkage and of the loads applied on an overlaid structure. Shrinkage is computed from the water loss associated, (1) to the hydration process, (2) to the water exchanges with the surrounding (mainly evaporation). The results faithfully represent the experimental relationship between measured shrinkage, applied loads and debonding. Apart from the boundaries, the Bernoulli principle appears as valid.

Féron [47] developed a mechanical model to determine the stresses in multilayer systems due to shrinkage, taking into consideration ground friction and chemo-thermo-hygro mechanical couplings in the materials. The water loss was calculated from diffusion laws by finite differences. The calculation of deformations and stresses in the multilayer system relies on the assumptions of (1) a perfectly bond between the layers, and (2) the respect of the Navier–Bernoulli principle. This model has been validated by comparative finite element simulations and successfully applied to the cases of industrial floors with Fibre Reinforced Concretes and repair layers.

HIPERPAV [48–51], although not a finite element code, takes into consideration thermo-hygro-mechanical phenomena by means of semi-empirical mathematical formulas from codes or recommendations, to predict the risk of cracking in overlays. The physical processes involved (thermal and moisture transport) are based on semi-empirical analytical solutions for current geometries.

The model from Rostasy et al. [52] is also numerical-analytical but includes very comprehensive descriptions of the physical and mechanical phenomena acting at early age and long term in composite structures. Although more dedicated to massive structures, walls on slabs for instance, this model is also applicable to overlays on substrates.

5.9 Conclusions

- Composite structural members are very common. They are however still often designed on an empirical basis.
- The notion of degree of restraint, associated to the degrees of freedom of a composite structural member provides a common background for the analysis of various practical applications among which are cement based overlays on existing concrete substrates, under restrained shrinkage.
- This notion initially derived for linear elastic overlay materials case can be extended to viscoelastic materials.
- This approach enables one to predict in a realistic way the stresses induced by restrained deformations and access the risks of cracking in the new layers.
- It can be further extended to three dimensional structures such as slabs, or to include the restraining effect of reinforcement bars in the substrate or in the new layer.

- For current cases of application (overlays on flexible substrates), in the linear-elastic case, the degree of restraint varies between 0.4 and 0.8 and can be calculated from simple geometrical and material parameters.
- Shear stresses at the interface of composite members can be accurately predicted, taking into consideration partial debonding.

5.10 Outlook for Future Research

- Experimental tests to determine accurately the strain profiles/distributions in various cases of member slenderness and combinations of materials.
- Extension to 3D (case of slabs) of analytical models taking into consideration the degree of restraint.
- Determination of design values of the time/rate dependent tensile strength.
- Implementation of non-linear viscoelastic models adapted for calculations with high stresses – high degrees of restraint.
- Relaxation tests under tensile loading at various pre-peak load levels, with temperature control of the specimen.

References

1. Silfwerbrand, J., Differential shrinkage in composite concrete beams of old concrete and a new-cast concrete overlay. Bulletin No. 144, Department of Structural Mechanics and Engineering, Royal Institute of Technology, Stockholm, Sweden, 149 pp., 1986 [in Swedish].
2. Silfwerbrand, J., Differential shrinkage in normal and high strength concrete overlays. *Nordic Concrete Research*, **19**, 55–68, 1996.
3. Bernard, O., Comportement à long terme des éléments de structure formés de bétons d'âges différents. Doctoral Thesis, Swiss Federal Institute of Technology No. 228, Lausanne, Switzerland, 2000. (in French).
4. Sadouki, H., Simulation et analyse numérique du comportement mécanique de structures composites, Doctoral Thesis, Swiss Federal Institute of Technology No. 676, Lausanne, Switzerland, 1987 [in French].
5. Birkeland H.W., Differential shrinkage in composite beams. *Journal of the American Concrete Institute*, 1123–1136, May 1960.
6. Silfwerbrand, J., Stresses and strains in composite concrete beams subjected to differential shrinkage, *ACI Structural Journal*, **94**(4), 347–353, 1997.
7. Alonso Junghans, M.T., Zur Risssicherheit zementgebundener dehnungsbehinderter Schichten unter Berücksichtigung von Dauereinflüssen. Ph.D. Thesis, Universität Hamburg-Harburg, Shaker Verlag, Aachen, Germany, 1997 [in German].
8. Denarié, E. and Silfwerbrand, J., Structural behaviour of bonded concrete overlays. In *Proceedings International RILEM Workshop on Bonded Concrete Overlays*, Stockholm, Sweden, June 7–8, 2004, edited by Swedish Cement and Concrete Research Institute (CBI), RILEM PRO 43, 2004.
9. Hartl, G., Kraftverlauf in Beschichtungen, *Zement und Beton*, **28**(2), 45–51, 1983 [in German].
10. Hartl, G., Materialtechnologische Beurteilung von Verstärkungsmassnahmen, *Beton und Stahlbetonbau* **95**(12), 707–712, 2000 [in German].

11. Beushausen, H., Long-term performance of bonded concrete overlays subjected to differential shrinkage, Ph.D. Thesis, University of Capetown, South Africa, 2005.
12. Beushausen, H. and Alexander, M., Bonded concrete overlays subjected to differential shrinkage – An analytical model based on localized strain and stress. In *Proceedings International RILEM Workshop on Bonded Concrete Overlays*, Stockholm, Sweden, June 7–8, 2004, edited by Swedish Cement and Concrete Research Institute (CBI), RILEM PRO 43, 2004.
13. Beushausen, H. and Alexander, M., Spannungen durch Verformungsbehinderung in gebundenen Aufbetonen, *Beton und Stahlbetonbau* **101**(6), 394–401, 2006 [in German].
14. Beushausen, H. and Alexander, M., Localised strain and stress in bonded concrete overlays subjected to differential shrinkage, *Materials and Structures*, **40**(2), 189–199, 2007.
15. Jonasson, J.-E., Computer program for non-linear computations in concrete with regard to shrinkage, creep, and temperature. CBI Report No. 7:77, Swedish Cement and Concrete Research Institute, Stockholm, 161 pp., 1977 [in Swedish].
16. Haardt, P., Zementgebundene und kunststoffvergütete Beschichtungen auf Beton, Dissertation, Heft 13, TU Karlsruhe, 1991 [in German].
17. Silfwerbrand, J., Concrete overlays, Report No. 10, 3rd Edition, Chair of Structural Mechanics and Engineering, Department of Structural Engineering, Royal Institute of Technology, Stockholm, Sweden, 65 pp., 1997 [in Swedish].
18. Silfwerbrand, J., Effects of differential shrinkage, creep and properties of the contact surface on the strength of composite concrete slabs of old and new concrete. Bulletin No. 147, Department of Structural Mechanics and Engineering, Royal Institute of Technology, Stockholm, Sweden, 131 pp., 1987 [in Swedish].
19. Beushausen, H. and Alexander, M., Failure mechanisms and tensile relaxation of bonded concrete overlays subjected to differential shrinkage, *Cement and Concrete Research*, **36**, 1908–1914, 2007.
20. Huet, C., Adaptation d'un algorithme de Bazant au calcul des multilames visco-élastiques vieillissants, *Materials and Structures*, **13**(74), 91–98, 1980 [in French].
21. Siestrunck, R., Lamer, P., Huet, C., and Alviset, L., Action de l'humidité sur la céramique envisagée dans le cadre de l'association béton-céramique. In *Communication C.T.T.B. au Symposium Rilem/C.I.B.*, Helsinki, 1965 [in French].
22. Gutsch, A. and Rostásy, F.S., Young concrete under high tensile stresses-creep relaxation and cracking. In *Proceedings RILEM Symposium on Thermal Cracking in Concrete at Early Ages*, R. Springenschmidt (Ed.), Chapman & Hall, London, pp. 111–116, 1995.
23. Horimoto, H. and Koyanagi, W., Estimation of stress relaxation in concrete at early ages. In *Proceedings RILEM Symposium on Thermal Cracking in Concrete at Early Ages*, R. Springenschmidt (Ed.), Chapman & Hall, London, pp. 95–102, 1995.
24. Denarié, E., Cécot, C. and Huet, C., Characterization of creep and crack growth interactions in the fracture behaviour of concrete, *Cement and Concrete Research*, **36**(3), 571–575, 2006.
25. FIP, Federation Internationale de la Precontrainte, Structural Effects of Time-Dependent Behaviour of Concrete, Guide to Good Practice, FIP, Wexham Springs, 1982.
26. Silfwerbrand, J., Composite action between partially chipped concrete bridge deck and overlay. Beam tests. Bulletin No. 142, Department of Structural Mechanics and Engineering, Royal Institute of Technology, Stockholm, Sweden, 72 pp., 1984 [in Swedish].
27. Chausson, H., La durabilité des rechargements minces adhérents en béton renforcé de fibres métalliques. Doctoral Thesis, No. 2708, Université Paul Sabatier/LMDC, Toulouse, France, 1997 [in French].
28. RILEM TC-119-TCE, Avoidance of thermal cracking in concrete at early ages – TCE3: Testing of the cracking tendency of concrete at early ages using the cracking frame test, *Materials and Structures*, **30**, 461–464, 1997.
29. Kovler, K., Testing system for determining the mechanical behaviour of early age concrete under restrained and free shrinkage, *Materials and Structures*, **27**, 324–330, 1994.
30. Bjontegaard, O., Thermal dilatation and autogenous deformation as driving forces to self-induced stresses in high performance concrete, Doctoral Thesis, Trondheim, Norway 1999.

31. Pigeon, M., Toma, G., Delagrave, A., Bissonnette, B., Marchand, J., and Prince, J.C., Equipement for the analysis of the behaviour of concrete under restrained shrinkage at early ages, *Magazine of Concrete Research*, **52**(4), 497–502, 2000.

32. Altoubat, S.A. and Lange, D.A., Tensile basic creep: Measurements and behavior at early age, *ACI Materials Journal*, **95**(5), 386–393, 2001.

33. Bentur, A. and Kovler, K., Evaluation of early-age cracking characteristics in cementitious systems, *Materials and Structures*, **36**, 183–190, 2003.

34. Charron, J-P., Contribution à l'étude du comportement au jeune âge des matériaux cimentaires en conditions des déformations libre et restreinte. Ph.D. Thesis, University Laval, Quebec, Canada, 2003 [in French].

35. Sule, M., Effect of reinforcement on early-age cracking in high strength concrete, Ph.D. Thesis, TU Delft, 2003.

36. Kamen, A., Denarié, E., and Brühwiler E., Mechanical behaviour of ultra high performance fibre reinforced concretes (UHPFRC) at early age, and under restraint. In *Proceedings CONCREEP 7*, G. Pijaudier-Cabot, B. Gérard, and P. Acker (Eds.), Hermès Publishing, pp. 591–596, 2005.

37. Martinola, G., Rissbildung und Ablösung zementgebundener Beschichtungen auf Beton. Dissertation. ETH Zürich, Switzerland, No. 13520, 2000 [in German].

38. Bernard, O. and Brühwiler, E., The influence of reinforcement in the new layer on hygral cracking in composite structural elements, *Materials and Structures*, **36**, 118–126, 2003.

39. Bernard, O. and Brühwiler, E., Influence of autogenous shrinkage on early age behaviour of structural elements consisting of concretes of different ages, *Materials and Structures*, **35**, 550–556, 2002.

40. Roelfstra, P.E., Salet, A.M., and Kuiks, J.E., Defining and application of stress-analysis-based temperature difference limits to prevent early-age cracking in concrete structures. In *Proceedings No. 25 of the International RILEM Symposium: Thermal Cracking in Concrete at Early Age*, Münich, pp. 273–280, 1994.

41. Habel, K., Structural behaviour of elements combining ultra-high performance fibre reinforced concretes (UHPFRC) and reinforced concrete. Doctoral Tthesis No. 3036, Ecole Polytechnique Fédérale de Lausanne, Lausanne, Switzerland, 2004.

42. Habel, K., Denarié, E., and Brühwiler, E., Time dependent behaviour of elements combining UHPFRC and concrete, *Materials and Structures*, **39**, 557–569, 2006.

43. Martinola, G., Sadouki, H., and Wittmann, F.H., Numerical model for minimizing the risk of damage in repair system, *ASCE Journal of Materials in Civil Engineering*, 121–129, 2001.

44. DIANA, User's manual-release 6.1, Diana Analysis BV, Delft, the Netherlands, 1996.

45. Laurence, O., La fissuration due au retrait restreint dans les réparations minces en béton: Apports combinés de l'expérimentation et de la modélisation, Doctoral Thesis, Ecole Nationale des Ponts et Chaussées, Paris, France, 2001 [in French].

46. Tran, Q.T., Toumi, A., and Granju, J-L., Experimental and numerical investigation of the debonding interface between an old concrete and an overlay, *Materials and Structures*, **39**(3), 379–389, 2006.

47. Féron, C., Etude des mécanismes de generation de contraintes et de fissuration par retrait gêné dans les structures à base de matériaux cimentaires, Doctoral Thesis, Institut National des Sciences Appliqués de Lyon, No. 2002 ISAL 0025, 2002 [in French].

48. McCullough, B.F. and Rasmussen, R.O., Fast-track paving: Concrete temperature control and traffic opening criteria for bonded concrete overlays. Volume I: Final report, Federal Highway Administration Report FHWA-RD-98-167, Washington, USA, 1999.

49. McCullough, B.F. and Rasmussen, R.O., Fast-track paving: Concrete Temperature Control and Traffic Opening Criteria for Bonded Concrete Overlays Volume II: HIPERPAV User's Manual, Federal Highway Administration Report FHWA-RD-98-168, Washington, USA, 1999.

50. Rasmussen, R.O., et al., Fast-track paving: Concrete temperature control and traffic opening criteria for bonded concrete overlays. Volume III: Addendum to the HIPERPAV user's manual, Federal Highway Administration Report FHWA-RD-99-200, Washington, 1999.

51. Rasmussen, R.O., McCullough, B.F., and Weissmann, J., Development of a bonded concrete overlay computer-aided design system, Research Report No. 2911-1, Centre for Transportation Research, The University of Texas at Austin, 1995.
52. Rostasy, F.S., Kraus, M., and Budelmann, H., Planungswerkzeug zur Kontrolle der frühen Rissbildung in massigen Betonbauteilen – Teil 5: Behinderung und Zwang, *Bautechnik* **79**(11), 778–789, 2002 [in German].

Chapter 6
Debonding

A. Turatsinze, H. Beushausen, R. Gagné, J.-L. Granju, J. Silfwerbrand
and R. Walter

Abstract This chapter focuses on the causes and the mechanisms of debonding of cement-based material overlays. Additionally, it describes how debonding affects durability of the composite structure. Methods for monitoring are described. The debonding mechanism is discussed, and debonding modeling is described. The role of reinforcement in debonding is important and is discussed. Crack propagation and crack opening affects debonding and is described. There are special overlays including those bonded to steel and overlays that are anchored to the substrate with metal anchors and they are discussed. The effect of boundaries and joints is discussed.

6.1 Introduction

This chapter focuses on the causes and the mechanisms of debonding of cement-based material overlays. Additionally, it describes how debonding affects durability of the composite structure. As described by Carter et al. [1], the term debonding implies, that a bond had previously developed between the repair layer and the substrate. The bond strength is one of the main parameter governing debonding. The development of bond depends on several factors, see, e.g., [2–7]. These factors are

A. Turatsinze · J.-L. Granju
Laboratoire Matériaux et Durabilité des Constructions (LMDC), UPS-INSA, Toulouse, France

H. Beushausen
Department of Civil Engineering, University of Cape Town, Cape Town, South Africa

R. Gagné
Département de Génie Civil, Faculté de Génie, Sherbrooke (QC), Canada

J. Silfwerbrand
Swedish Cement and Concrete Research Institute (CBI), Stockholm, Sweden

R. Walter
Department of Civil Engineering, Technical University of Denmark (DTU), Lyngby, Denmark

not discussed herein (it is dealt with in Chapter 4) but will remain implicit in this chapter.

6.2 Impact of Debonding

Substrate-overlay composites should work monolithically, and the bond at the interface ensures the continuity of deformations between the base and the overlay. In these conditions, the durability of a repair relies not only on the durability of the material used, but also on the durability of its bond with the base.

6.3 Brief Summary of Debonding

A holistic model presented by Vaysburd et al. [8] and Emmons et al. [9] shows that whatever the original cause of the repair failure is, cracking is involved. According to Carter et al. [1], there are many possible causes of debonding initiation also for properly placed overlays. They include adverse ambient weather conditions during construction, improper wet curing, excessive vehicle impact due to wear surface roughness, formation of cracks allowing water access and the subsequent frost, vibrations during overlay construction and minor bonding defect. In a mechanical respect, it is well accepted that there are two classes of debonding causes: the effects of the external mechanical loading and the effects of the different length changes of the overlay and of its substrate. Additionally it is known that debonding initiates preferentially from discontinuities of the overlay: boundaries, cracks and joints. Here, the straining of the overlay induces concentrations of built-in stresses that act to debond. Although, the debonding mechanism is related to a mixed mode cracking [10, 11], mode I cracking dominates at debonding initiation. The vulnerability of boundary locations, joints and cracks, is inherent in their deficiency in stress transfers. Such a weakness cannot be corrected for the boundary locations but it can be attenuated for cracks and partial depth sawed joints thanks to overlay reinforcement. Reinforcement can be applied using traditional reinforcements such as steel fabrics and steel bars or by fibres in the ready mix concrete. Laboratory tests, field trials and numerical simulations lead to the conclusion that these various types of reinforcement efficiently improve durability of cement-based overlay (see Section 6.7).

6.4 Methods for Monitoring Debonding

There are several methods to detect debonding both for in situ and laboratory evaluations. In both cases non-destructive and semi-destructive methods can be implemented. These methods are briefly described below.

6.4.1 Non-destructive Methods

- Visual inspection is mainly a laboratory method because the eye must have access to the side edge of the overlaid structure. The method is subjective, and the help of magnifying devices significantly improves accuracy. A video microscope is especially efficient.
- Chain drag – The method is rudimentary but gives a qualitative assessment in good agreement with results from more sophisticated methods [12–14]. It consists of dragging a chain across the overlay surface while listening for a change in acoustic response. It is a rapid and inexpensive method.
- Infrared thermography [13–15]. Delaminated areas appear as areas of different temperature. This technique offers some advantages:

  - no direct contact required between the camera and the object under investigation;
  - monitoring of quite large areas with an accuracy which can reach ±0.08°C; and
  - the required equipment is lightweight and can be easily transported.

  Drawbacks are:

  - emissivity varies from material to material and with the brightness of different objects within an area;
  - high sensitivity to environmental conditions; and
  - the results are influenced by the time of the day and whether the object is subjected to heating or to cooling.

- Electromagnetic methods. Typically Ground Penetrating Radar (GPR): it is a standardised test (ASTM D6087) and a widely documented method to detect delaminations [13].
- Acoustic methods: acoustic emission, ultrasonic pulse velocity and impact-echo [14, 16–18]. These methods are generally reputed to be more expensive and time consuming. In return, they are quantitative methods.

6.4.2 Semi-destructive Methods

The widest spread ones are:

- the pull-off test (presented in Chapter 4); and
- coring and visual plus mechanical investigation of the core.

6.4.3 Laboratory Tests

In addition to the above listed methods, in laboratory conditions, accessibility of the boundary locations provides additional information. Typically LVDTs or strain gages spanning the interface are placed and will be activated when debonding reaches their location [19]. Based on their low cost and high reliability, LVDT sensors are often recommended [20]. Farhat [21] successfully used a set of breaking electrical conductive links spanning the overlay-substrate interface. Dynamic measurements giving the structure's eigen frequencies have also been used. Increasing cracking would lead to decreasing stiffness and frequencies [22].

6.5 Debonding Mechanism

Debonding of an overlay is considered to result from one or the superposition of the two causes sketched in Figure 6.1, respectively, designated as "due to flexural effects" (Figure 6.1a) and "due to differential length change" (Figure 6.1b). In situ, these two causes of debonding superimpose their effects.

In both cases, debonding results from a full depth separation of the overlay, e.g. a crack, disabling the continuity of stress transmission within the volume of the overlay. Emmons et al. [9] and Bernard [24] also pointed out the prominent role of cracking in the debonding mechanism.

At the crack location, sharp built-in peak stresses develop at the interface, which subsequently causes debonding [25–28]. Reinforcement of the overlay, which enables force to be transmitted through the crack, decreases the intensity of the mechanical discontinuity at the crack location. Consequently, it decreases the built-in peak stresses at the interface. It is through this process that reinforcement of the overlay, notably fibre reinforcement, acts to improve the durability of the bond to the substrate.

This is schematically shown in Figure 6.2 [29] in the example of a debonding due to loading stresses (where d is the distance from the loading point). It compares the case of: (1) an un-cracked overlay, (2) a neat cut through the overlay simulating a crack in a non-reinforced overlay (no force is transmitted through the crack), (3) a lower elastic modulus section (one-third of the modulus of the repair material) inserted in the overlay to simulate an overlay section containing a fibre reinforced crack. In the case of "fibre reinforcement", the peak stresses acting to debond are reduced by 75 to 80%.

In the case of steel reinforcement bars or welded fabric the same mechanism is involved. Steel bars and welded fabric can transmit much higher forces than fibre reinforcement, but they can only be used with thick overlays with respect to cover depth requirements. Another advantage of fibres is that, unlike steel bars, they present a reduced risk of corrosion [30].

Whatever the origin of debonding is, "mechanical" or "length changes", the basic cause of developed tensile stress, perpendicular to the interface, is the so-called

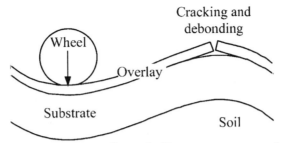

(a) Debonding due to flexural effects, consequence of the flexural straining of the structure by the applied loads, especially by live loads. The critical zone is apart from the applied loads, where the overlay is in tension. Debonding is possible because of the cracking of the overlay.

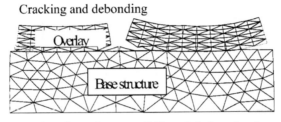

(b) Debonding due to differential length change, induced by the effects of the different length changes of the overlay and of the base, because of shrinkage or temperature changes. Debonding is a consequence of the cracking of the overlay. At each side of the crack, the peeling effect induces tensile stresses perpendicular to the interface.

Fig. 6.1 The two fundamental causes of debonding [23]: (a) due to flexural effects, (b) due to differential length change

peeling effect. It is the consequence of the unbalanced shear stresses at a cut of the bonded overlay (cracks, saw-cut joint or slab edge). It is easily illustrated by the peeling moment as presented in Figure 6.3. The peeling moment increases with the edge overlay thickness.

In the case of a cement-based overlay, because of its significant thickness (often more than 20 mm), the peaks of shear stress and of tensile stress are often of the same order of magnitude. Knowing that the tensile strength of an interface between cement-based materials is at the most 50% of its shear strength [19, 31–33], it is demonstrated that debonding is always initiated by tension perpendicular to the interface [19, 29, 34, 35].

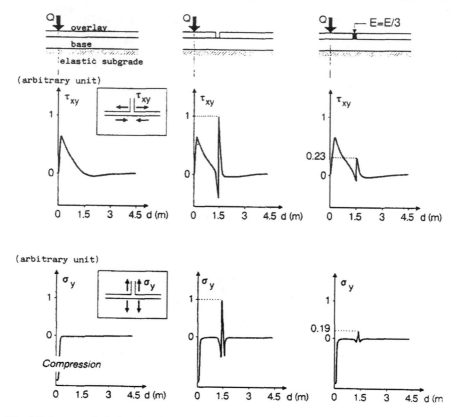

Fig. 6.2 Stress peaks induced at the interface by a crack and effect of fibre reinforcement, example of a debonding of mechanical origin [29]

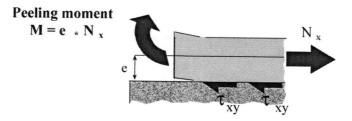

Fig. 6.3 Peeling effect [23]

In the case of debonding caused by length changes, a large amount of information is available, notably on the effect of restrained shrinkage [36–39]. Fowler et al. [26] presented thermally-induced stresses at the interface with a polymer concrete used as a material repair. It was demonstrated that debonding could occur due to length change alone, and the model confirmed that debonding preferentially starts at the overlay boundaries.

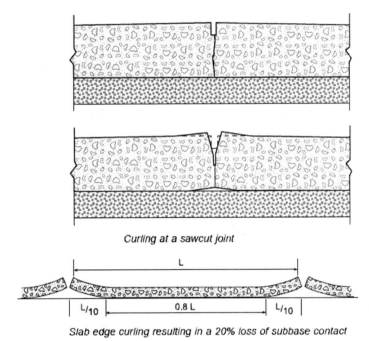

Curling at a sawcut joint

Slab edge curling resulting in a 20% loss of subbase contact

Fig. 6.4 Curling effect at cracks, saw-cut joint and slab edge [40]

Curling (Figure 6.4) due to non uniform shrinkage through the thickness of the overlay [40–42] increases the built-in tensile stress perpendicular to the interface. In the case of mechanical origin debonding, the trend of the overlay to not follow the curvature of the substrate, designated in the following by "geometrical effect" (Figure 6.1a) is also an aggravating factor.

6.6 Debonding Modeling

Numerical modeling offers exciting tools to investigate debonding mechanism. Debonding is modeled as a crack propagating along the substrate-overlay interface. It is now known that the strain softening (generally designated in the following as interlocking) is essential in modeling crack propagation. For the same reasons debonding modeling has to integrate the softening behaviour of the interface. Moreover, it must be taken into account that the debonding mechanism is characterized by mixed mode, mode I plus mode II. Recently Walter et al. [11] and Sabathier et al. [10, 43] proposed an improved model. They consider interaction between residual interfacial shear and tensile stresses. Breakage of bond in pure tension will necessarily, at some point, affect the shear interlocking, and vice versa. Tran et al. [44] completed this modeling. It shows that in the case of debonding of mechanical origin (bending of the overlaid structure with the overlay on the tension side) mode II can be neglected. The same authors also presented an interesting numerical analysis of factors influencing the overlay debonding taking account of shrinkage strains, calculated from the hydration development and water exchanges between the overlay and its surroundings, and including the fatigue damage of interlocking (in the cracks and along the debonded interfaces). In addition, effects of Young's modulus and of tensile strength of the overlay material, effects of type of the overlay reinforcing fibres (slipping or high bond fibres) and the influence of tensile strength of overlay-substrate interface were investigated. In particular Tran et al. [45] demonstrated that a low modulus of elasticity and high strength, two characteristics mutually exclusive, are suitable for durable thin cement-based overlays.

6.7 Role of Reinforcement of Overlays

The reinforcement can be provided by steel bars or welded fabric or by fibres. Although all these types of reinforcement act in similar ways, the literature mainly reports cases of fibre reinforcements.

6.7.1 In Situ Findings

One of the first and most convincing attempt to use fibre reinforced bonded overlays was made in Greene County (Iowa, USA) to repair concrete road pavements [46]. It illustrated the beneficial effect of steel fibre reinforcement of the overlay on the durability of the repair. The fibres were common steel wire fibres. The fibre content was very high, 90 kg/m^3. Paulsson et al. [47] reported a similar case, with success, where 75 kg/m^3 of "enlarged end" steel fibres were used as reinforcement.

Another typical example is the Belgian experience in resurfacing existing roads with concrete, which has been in use for about 20 years. Improved incorporation methods allowed for subsequent lower fibre content in thin bonded repairs, less than 150 mm thick [48].

Other results from various places confirmed the positive effect of metal fibre reinforcement of the overlay even at much smaller dosage, from 25 to 40 kg/m^3. A significant example is the case of the Highway 40 in Montreal, Canada. It has three lanes in each direction withstanding a traffic intensity of 30,000 vehicles per day. In 1986, an experimental repair was performed with the goal to compare the efficiency of different types of bonded overlays. On a total length of 134 meters, each investigated overlay system covered a section of about 20 meters in length. This experimental repair is interesting because it is thoroughly documented. Its conception, the preparation of the substrate, and the placing of the overlay were supervised by a team from Sherbrooke University (Quebec, Canada) which, further, ensured the monitoring of its behaviour over a 12-year period [12, 15, 49, 50]. The different

Table 6.1 Fibres used and 28-day compressive strengths of the mixes

Type of fibre	Steel fibres					
	A	B	A	B	C	
Dosage (kg/m^3)	22	24	34	34	34	0
28-day compressive strength f_c (MPa)	53.2	45.8	54.4	49.2	28.9	46.4

Fibre A: length: 50 mm, shape: hooked end, section: circular, diameter: 0.50 mm
Fibre B: length: 53.52 mm, shape: undulated, section: rectangular, equivalent diameter: 0.96 mm
Fibre C: length: 60 mm, shape: undulated, section: circular, diameter: 0.1 mm

types of overlay systems accounted for and compared included plain concrete (100 mm thick) and others (75 mm thick) reinforced with 22 to 34 kg/m^3 of steel fibres. Three types of steel fibres were tested: hooked ended, and undulated, of circular or rectangular cross section, all 50 mm long. The effect of attachment of the overlay to the base pavement by steel nails was also investigated. These were 37.5 mm long and previously driven to half of their length into the old pavement. No nail was driven in the central lane; the two other lanes were nailed, one with a nail 300 mm long, and the other with a nail 450 mm long. The interface was prepared by sand jetting to remove oil stains, other impurities and all lose material. In the case of the plain concrete overlay, 25 mm of the old concrete pavement was removed by scarification. All the overlays were bonded with a thin cement grout (even if as it is discussed in Chapter 4 the actual influence of a such bonding agent is still a controversial topic). A sketch of the different test sections is shown in Figure 6.5.

The repair materials were concrete with a maximum aggregate size of 20 mm and a water-cement ratio of 0.38 (160/420). The 28-day compressive strengths of the mixes are given in Table 6.1.

Cracking detection was achieved by visual inspection. For improved detection, observations were carried out on a slightly moist repair surface. The crack growth, quantified by the total length of observed cracks per unit length of lane, is presented in Figure 6.6. In the plain concrete overlays a sharp rate of crack growth was observed associated with a quick deterioration of the pavement: cracks induced breaking in many sections and led to debonding of the overlay, and within less than two years the overlay had to be replaced. On the other hand, in the fibre reinforced sections, the crack growth was much slower and almost vanished after the first year. No significant difference was seen between the different types and dosages (from 22 to 34 kg/m^3) of steel fibres. After 12 years of use, the fibre reinforced repairs were still sound, with limited cracking and no debonding. The current cracks remained narrow and limited in numbers. The majority of cracks were hairline cracks and developed to less than half the depth of the overlay. However, the reflection of the cracks or other discontinuities of the substrate structure could not be prevented. Their opening was reduced when fibre reinforcement was used. As illustrated in Figure 6.7, the presence of connecting nails had an insignificant effect on the crack growth.

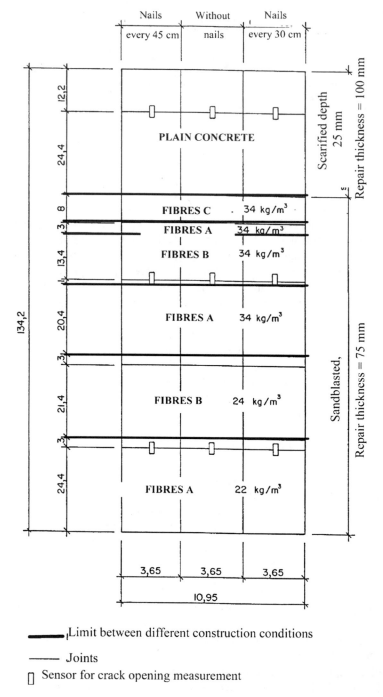

Fig. 6.5 In situ experimental rehabilitation – details of different construction conditions [12, 15, 49, 50]

length of observed cracks
per unit length of lane (m/m)

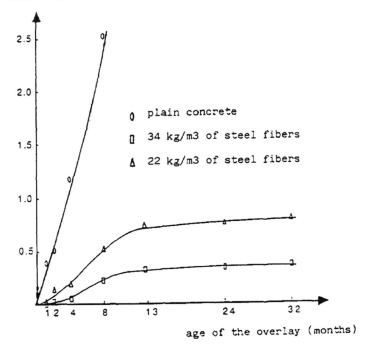

Fig. 6.6 Crack growth versus time, effect of fibre reinforcement [12, 15, 49, 50]

For the fibre dosages used (less than 40 kg/m^3) shrinkage and thermal length changes were not significant. Consequently, the explanation of the beneficial effect of the fibre reinforcement of the overlay cannot be found in lower length changes between the overlay and the substrate. The beneficial effect of the fibres actually must come from their ability to transmit significant through-crack stresses and to restrain cracking.

6.7.2 Need to Distinguish between "First Monotonic Loading" and "Shrinkage-Pre-cracking Plus Fatigue"

6.7.2.1 First Monotonic Loading

Most of the laboratory tests are related to first monotonic loading (or first monotonic straining) cases.

A thorough investigation carried out in Toulouse (France) over ten years demonstrated the correlation between cracking and debonding and confirmed the benefi-

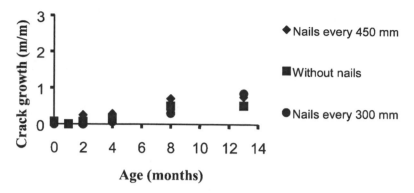

Fig. 6.7 Crack growth versus time, effect of connecting nails: case of a concrete reinforced with 22 kg/m³ [12, 15, 49, 50]

cial effect of fibre reinforcement of the overlay [21, 44, 45, 51–55]. The case of debonding of mechanical origin was approached by three-point flexure tests with the overlay on the tension face.

To limit scatter of results, the concrete substrates were replaced by steel hollow shapes designed and treated to simulate concrete substrates. Their inertias were similar to that of concrete substrates of same sizes. The interface with the overlay was treated in order to have the same characteristics as a concrete-to-concrete interface: the surface of the base was milled to a rough finish, and cycles of casting, curing and debonding of a cement-based overlay were repeated until the metal was covered by a thin, uniform adhesive layer of mortar (for more details, see [55]). The interface bond shear strength was in the order of twice its tensile strength. Actually, with a 50 MPa 28day compressive strength mortar as overlay, measured bond tensile strength and bond shear strength were about 0.5 and 1 MPa, respectively. The test samples were all 100 mm wide with a flexural span of 430 mm. The height of the substrates varied from 50 to 150 mm and the thickness of the overlays (tensile side) from 20 to 60 mm. The curvature of the structure was characterised by the deflection measured at mid-span. Depending on the test series, debonding initiation and propagation were detected by three LVDTs or by five electrical strain gages or a set of breaking electrical conductive links spanning the overlay-substrate interface. Visual observation with a magnifying device (×25 at least) completed the information.

Two types of fibres were investigated: usual hooked steel wire fibres and flexible ribbon-shaped amorphous metal fibres, both types were 30 mm in length. The first ones were designed to slip inside the concrete matrix and not to break and were designated as slipping fibres; they provided a long but relatively low post-crack strength plateau. The second ones, characterized by a high bond with the concrete matrix, provided a high but short post-crack strength plateau.

Such tests, illustrated in Figure 6.8, demonstrated the beneficial effect of a fibre reinforcement of the overlay. As long as there is no debonding, the devices spanning the interface measures compression. When debonding occurs, tension is indicated,

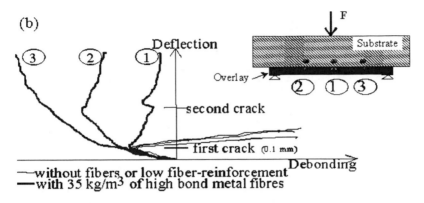

Fig. 6.8 Debonding of mechanical origin: an example of test results

rightwards on the plot of Figure 6.8. It can be seen that, without fibres, debonding occurs at the same deflection as the first crack and instantly propagates over all the investigated length along the interface. With 35 kg/m^3 of high bond metal fibres, the first crack initiated debonding but there was no propagation. A second crack, which opened in the vicinity of LVDT 2, initiated another debonding at this location, but still with no significant sign of propagation. Obviously, the fibre reinforcement of the overlay had a beneficial effect.

In the case of a first monotonic loading, 35 kg/m^3 of high bond fibres were necessary to bring a significant delay of debonding. With slipping steel fibres, 80 kg/m^3 were necessary for a similar result.

Debonding is spotted by the signals of three LVDTs. The diagram indicates the evolution of signals of the transducers 1, 2 and 3 with beam deflection, debonding is evidenced by the rightwards move of the plot [52, 54].

More detailed information is provided by the modeling of these tests, by Sabathier [43]. A sample of the results presented in Figure 6.9 clearly illustrates that fibre reinforcement by 35 kg/m^3 of high bond metal fibres of the overlay significantly slows down the rate of debonding propagation. But the beginning of the curves shows that, in the present case of a first monotonic loading, this beneficial effect is not clearly visible on the debonding initiation nor on the very first phase of debonding propagation.

Bernard [24] and Habel et al. [56] carried out a study focusing on the case of hybrid reinforced concrete elements. It combined experiments and modeling. For this purpose, the thickness of the overlays exceeded 100 mm. Given a concrete substrate 150 mm thick, different overlay thicknesses were investigated in the range of 70 to 170 mm. The focus was on debonding initiation, rather than the extent of debonding propagation. Concerning the role of fibre reinforcement, the results presented in Figure 6.10 are in agreement with those presented above. Reinforcement with steel bars or welded fabric concentrates higher crack-bridging forces, thus a more significant effect is present.

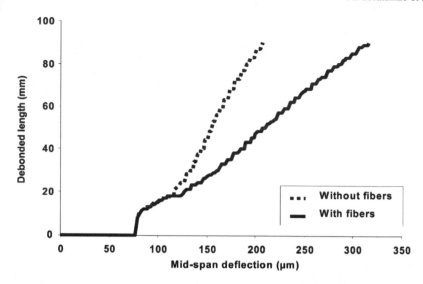

Fig. 6.9 Effect of a fibre reinforcement of 35 kg/m^3 of high bond metal fibres on debonding initiation and propagation: a 20-mm thick overlay on a 50-mm deep substrate, modelling by Sabathier [43] of Chausson's tests [54]

From these results, the authors concluded that it is better to improve the bond tensile strength or the overlay tensile strength rather than to add fibres. Moreover, they demonstrated that reinforcement content required in the new layer to avoid initiation of debonding is a function of the ratio of thickness of the new to old concrete. With a high ratio, it is more difficult to control debonding, and the interfacial adherence strength must be high to ensure monolithic behaviour.

These sets of investigations in first monotonic loading condition confirm that 80 kg/m^3 of hooked ended steel fibres have a small effect on the structural behaviour and have a minor effect to delay debonding initiation.

6.7.2.2 Shrinkage-Pre-cracking and Fatigue Loading

In accordance with some results from first monotonic loading presented in previous sections, a minimum amount of 80 kg/m^3 of common steel fibres is needed to observe an improvement, which is in the range of three times more than the 22 to 27 kg/m^3, which demonstrated a total efficiency in the Montreal repair.

Farhat et al. [57], Turatsinze et al. [58], and the analysis of the last results of Sabathier and Turatsinze [10, 55] explain this very large difference by a series of hypothesis on the cumulative effects of restrained shrinkage, inducing early cracking, and of fatigue, damaging the concrete interlock in the cracks of the overlay. Indeed, in real slabs on soil or pavements the maximum expected curvature placing the overlay in tension is much less than the curvature necessary to initiate cracking

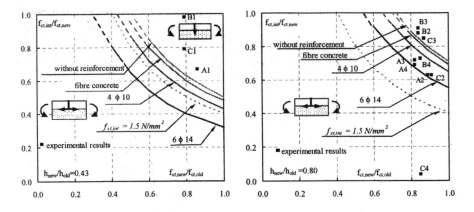

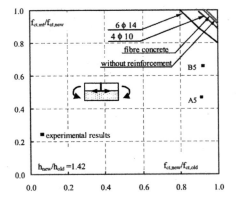

where: Thickness of the new/old concrete: h_{new}/h_{old}

Tensile strength of the new/old concrete: $f_{ct,new}/f_{ct,old}$

Tensile strength perpendicular to the new-old concrete interface: $f_{ct,\,int}$

Content of reinforcement in repair layer:

Fibre reinforcement 80 kg/m^3 hooked end steel fibres

Diameter   10 mm with a spacing of the rebars = 150 mm

Diameter   14 mm with a spacing of the rebars = 80 mm

Fig. 6.10 Required properties of the interface to avoid delamination: area below the curves indicates partial or total delamination will occur [56]

in the case of a first mechanical monotonic loading, and it is the effect of restrained shrinkage which makes cracking possible, designated as "shrinkage-pre-cracking". In addition, the fatigue type loading and vibrations caused by the vehicular wheel loads accelerates cracking initiation. At this low level of curvature, a low quantity of fibres can have a visible effect. After cracks have been initiated, still at this low level of curvature, fatigue causes their gradual propagation by gradual damaging of the interlocking in the cracks. Considering interlocking by two components, the

Table 6.2 Main mechanical characteristics of the repair materials

	M0	M1	M2
f_c (MPa)	60	55	51
f_t (MPa)	3.5	3.4	3.4
σ_r (MPa)	–	1.3	1.6
w_c (mm)	–	several	0.2
w_l (mm)	–	several	1.0

one provided by the cementitious matrix and the one brought by the fibres bridging the crack, the portion coming from the fibre reinforcement is much less affected by fatigue than the one coming from the cementitious matrix [59, 60]. This asymmetry in the cracking, in the case of fatigue, indicates that the global influence of the fibre reinforcement is more visible. That is schematically illustrated in Figure 6.11.

Farhat et al. [57] and Turatsinze et al. [58] provided an experimental confirmation of this hypothesis. The experimental set-up, similar to the one presented in Figure 6.8 is detailed in Figure 6.12. It was completed using a pre-notch when casting located at mid-span of the overlay. Due to shrinkage, this notch initiated a crack in the overlay, at zero curvature, before the time of testing (at the 7th day). Moreover, the notch located the crack and made possible to fit a LVDT sensor to monitor the crack opening.

The metal bases were 150 mm high and the overlay 60 mm thick. The samples were tested in sinusoidal fatigue loading at a frequency of 5 Hz between 0.1 and 9 kN. The maximum load (9 kN) corresponded to the lowest load necessary to propagate cracking from the tip of the notch when the overlay was protected against drying shrinkage. The associated curvature was significantly lower than the one imposed during the "first monotonic loading" type tests, performed on notched specimens. Three types of mortar overlays were investigated. Mix M0 without fibre, mix M1 reinforced with 40 kg/m^3 of 30 mm long hooked steel fibres (slipping fibres) and mix M2 reinforced with 20 kg/m^3 of 30 mm long high bond metal fibres. The main mechanical characteristics of the mixtures are given in Table 6.2, in terms of f_c and f_t, the compressive and tensile strength values respectively and of σ_r, w_c, w_l that characterise the post-crack load-bearing capacity of the mixture as indicated in Figure 6.13. These last three parameters are obtained from a tensile test on notched specimen.

The results are presented in Figure 6.14. They confirm that, in "shrinkage-pre-cracking plus fatigue" condition, fibre reinforcement significantly less than 80 kg/m^3 of slipping fibres can have a significant effect to delay debonding initiation and propagation. They also indicate that, although their short post-crack strength plateau, the high bond fibres remain fully efficient: 20 kg/m^3 of metal high bond fibres are more efficient than 40 kg/m^3 of metal slipping fibres.

Gagné et al. [61], Lemieux et al. [62] and Guindon [63] generated additional experimental data that confirm and clarify some of the debonding mechanisms previously described. They notably confirmed that, at load level representative of actual concrete slabs, debonding "usually does not start during the first monotonic load-

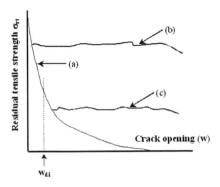

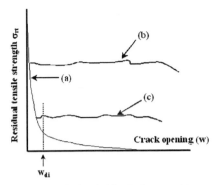

Case of a "first monotonic loading": debonding occured early and even 30 kg/m³ of fibres was not observed to have any effect.

After fatigue: the interlocking in the cementitious matrix is greatly damaged (it vanishes at thinner crack opening), the crack bridging capacity of the fibre reinforcement is affected but still remains significant. In such conditions, 30 kg/m³ of fibres is effective in delaying debonding.

Fig. 6.11 Schematic illustration of the effect of fatigue on the influence of a fibre reinforcement on debonding initiation; (a) without fibres; (b) with high fibre dosage ($\approx$ 80 kg/m³) of slipping steel fibres; (c) with usual fibre dosage ($\approx$ 30 kg/m³) of slipping steel fibres; w_{di}: crack opening corresponding to the debonding initiation

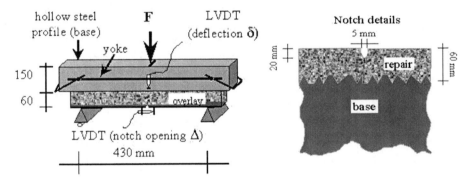

Fig. 6.12 Experimental set-up for "shrinkage-pre-cracking and fatigue" tests, details of the base-repair interface and of the repair notch [58]

ing but gradually develops under cyclic loading". It should be noted the effect of the interlocking damaging by fatigue. Their experimental program was designed to provide a better understanding of the structural behaviour and durability of thin bonded concrete overlays. An additional purpose was to provide technical recommendations in order to improve the technique's performance in practice. It involved testing 15 simply supported reinforced-concrete slab panels (3.3 m × 1.0 m × 0.2 m) designed to support actual service loads. Slab design, dimensions and steel reinforcement were designed to accurately simulate the characteristics of typical panels in bridge decks.

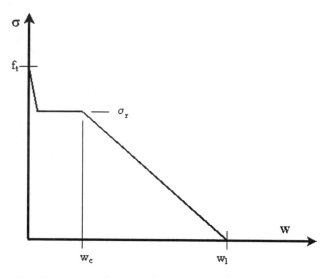

Fig. 6.13 Post-crack load-bearing capacity versus crack opening

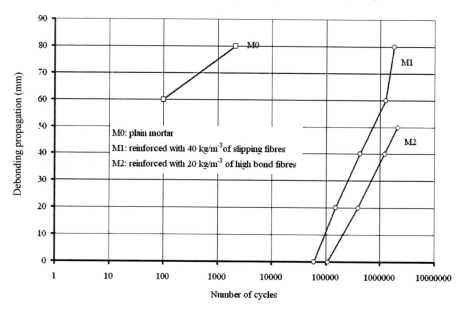

Fig. 6.14 Case of "shrinkage-pre-cracking plus fatigue", debonding propagation versus cycles number [21, 55, 57]

Some slab panels were hydro-demolished to a depth of 20 mm to simulate "shallow" deterioration (non-exposed reinforcing steel) while others were hydro-demolished to a depth of 95 mm to simulate "deep" degradations (exposed reinfor-

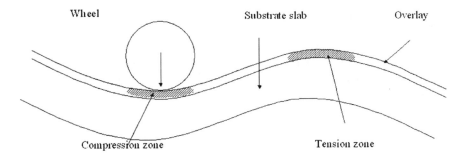

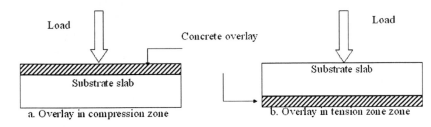

Fig. 6.15 Overlays location and loading mode [62]

cing steel). Five types of concretes were used to produce the overlays: 35 MPa, 35 MPa + 5% Latex, 35 MPa + steel fibres, 50 MPa, and 50 MPa + steel fibres.

All fibrous concretes contain 40 kg/m^3 of 60-mm hooked steel fibres. All slabs were subjected to cycling loading (up to 500,000 cycles) and monitored for load, displacement, flexural cracking and debonding (from observation of lateral faces). Figure 6.15 shows that flexural tests were designed to simulate either zones of positive (repair in compression) or negative moments (repair in tension).

Overall, no debonding was observed with the overlay located in the compression zone [62]. This absence of damage was mainly responsible for the good structural behaviour of this category of slab panels, regardless of the configuration type assessed. Results strongly suggest that it is possible, in the case of an overlay located in compression zones, to develop and maintain monolithic structural action in the substrate/overlay composite.

Some debonding occured for overlays located in tension zones. For this loading mode, debonding was influenced by interface location (above or below reinforcement), fibres, type of concrete and number of cycling loading. When debonding occurred, it usually did not start during the first monotonic loading but gradually developed under cycling loading. Abrupt debonding (first loading) only occurred with the two non-fibre-reinforced high-strength concrete overlays.

Overall, for overlays located in tension zones, the results indicated that, associated to their cracking, the debonding trend increased with the curvature imposed to the overlaid structure. Once the flexural cracks reached the interface, tensile stresses developed at the interface between the two concrete layers because of peeling and because the overlay resisted the curvature. The behaviour was similar to that of a plane rigid concrete plate stuck to a slightly curved surface; it was the "geometrical effect". In case of debonding of mechanical origin, peeling and geometrical effect are tightly linked. Tensile stresses developed perpendicular to the plane of the interface along plate edges. As the curvature of the support increased, the tensile stresses generated at the edges of the plate increased (Figure 6.16).

These tensile stresses were probably not large enough to initiate debonding under the first monotonic loading. However, the results showed that the level of stress under cyclic loading was sufficient to generate fatigue debonding and edge curling due to differential curvature between the substrate and the overlay. According to these mechanisms, thick or highly rigid overlay concretes will form rigid "blocks" between flexural cracks. Each block strongly resists the curvature imparted by the substrate slab and, therefore, develops high tensile stresses perpendicular to the plane of interface along block edges. This mechanism was experimentally confirmed by the sudden debonding of high-strength non-fibre-reinforced overlays during the first monotonic loading of the slabs [62, 63].

The use of relatively low fibre content (40 kg/m^3 anchored steel fibres) contributed to reduce the risk of interface debonding. Experimental results indicated that overlays containing fibres had a higher number of finer flexural cracks [63]. The bridging action of fibres and formation of smaller blocks in the fibre-reinforced overlays contributed to reduce the magnitude of tensile stresses perpendicular to the interface.

Interface debonding was reduced for thick overlays containing reinforcing bars. The action of the reinforcing bars seems to have more uniformly distributed tensile stresses at the interface. The "blocks" are thus smaller and, consequently, interface debonding is less severe [62].

In general, the results showed that the use of a bonded concrete overlay in a tension zone does not compromise the structural capacity of a reinforced concrete slab when the repair is carried out with appropriate techniques. Damage at the overlay/substrate interface has little effect on the structural capacity of the repaired slab, however, because the reinforcing steel under tension provides greater resistance to the generated tensile stresses [62].

6.7.2.3 Conclusion

Shrinkage precracking plus fatigue, representative of actual in-situ conditions, drastically change the levels of the relevant parameters involved in debonding initiation and propagation. Any reliable attempt of behaviour prediction must take that into account.

First, shrinkage (or more generally differential length changes of the overlay and the substrate) induces built-in stresses that make cracking occur at very low, even nil, curvature levels. Second, fatigue progressively damages the available interlocking

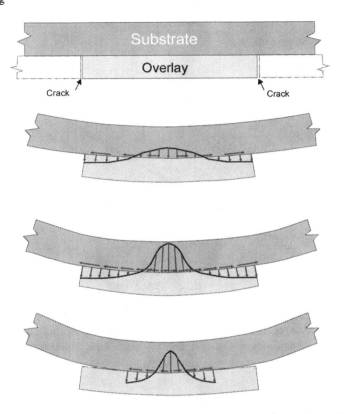

Fig. 6.16 Evolution of tensile stress imposed by the curvature of the substrate [63]

in the crack and along the debonded area of interface. Over the long term, only the interlocking provided by a reinforcement of the overlay will be significant, and consequently it will then play a major role.

On this basis, for a realistic calculation including fatigue effects, a conservative solution would be to neglect all interlocking excepted that provided by overlay reinforcement. Figure 6.17 presents the results of Sabathier [43], modeling Chausson's tests [54]. They indicate that the calculated crack opening at debonding initiation is less than 10 µm and that, for a debonded length of 180 mm (90 mm each side of the initiating crack), the crack opening remains thinner than 120 µm.

6.8 Crack Propagation and Crack Opening

Fibres and other reinforcements act to delay debonding by their ability to restrain crack opening. What crack opening width is the limit to prevent debonding?

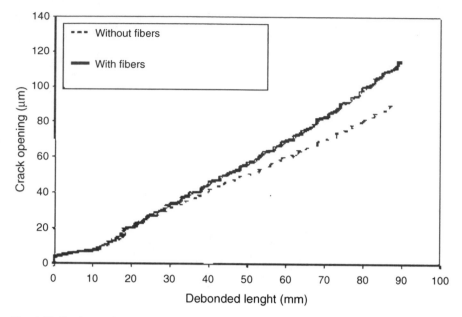

Fig. 6.17 Crack opening versus debonding propagation: modeling of Chausson's tests [54] by Sabathier [43] (a 20-mm thick overlay on a 50-mm deep substrate; with fibres: 35 kg/m^3 of high bond metal fibres)

6.8.1 First Monotonic Loading

In this straining condition, the cracks propagate quickly through the overlay, and their opening at debonding initiation is very small. Meanwhile, it is difficult to get an accurate experimental measurement. To investigate this initial phase, modeling is of great help. To investigate the subsequent phase, debonding propagation, experimental measurements and modeling are complementary.

Figure 6.17 presents the results of Sabathier [43], modeling Chausson's tests [54]. They indicate that the calculated crack opening width at debonding initiation is less than 10 μm and that, for a debonded length of 180 mm (90 mm each side of the initiating crack), the crack opening remains thinner than 120 μm.

Tests carried out by Farhat [21, 57, 58], with the test fitting already described in Section 6.7.2 but in monotonic loading, brought a complementary experimental information. They were different from the above calculation by the shrinkage-pre-cracking and a lower fibre dosage (20 kg/m^3). The consequence of pre-cracking was larger crack openings. As indicated in Figure 6.18, crack openings ranged from 0.05 mm at debonding initiation to about 0.2 mm for a total debonded length of 160 mm (80 mm each side of the initiating crack). It confirms that to prevent or limit debonding, the reinforcement must act as soon as the cracks appear and that beyond a 0.2-mm crack opening it is too late.

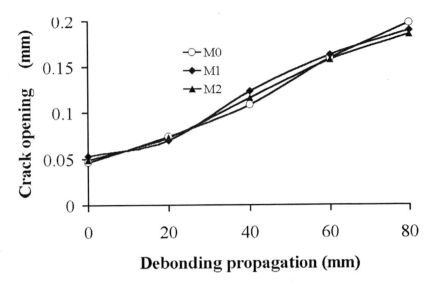

Fig. 6.18 Crack opening versus debonded length each side of the initiating point: case of mono-tonic loading after shrinkage pre-cracking (M0: plain mortar, M1: reinforced with 40 kg/m^3 of slipping fibres and M2: reinforced with 20 kg/m^3 of high bond fibres) [21, 58]

6.8.2 Shrinkage-Pre-Cracking and Fatigue Loading

The monitoring of the Montreal motorway repair concluded that, in the case of the fibre reinforced repairs, debonding was prevented and, except the cracks reflected from the substrate, only very thin hairline cracks were observed. Moreover, most of them had not propagated through the full depth of the overlay.

This trend is confirmed by the tests performed by Farhat [21], Turatsinze et al. [55] and Farhat et al. [57] already presented in Section 6.7.2. Their results demonstrated that, in in-situ condition (shrinkage-pre-cracking and fatigue at a low curvature level), the crack actually propagated gradually through the overlay. This is illustrated in Figure 6.19. The results showed a first phase of practically instant-aneous propagation until a depth related to the maximum imposed curvature. After-wards, in the case of fibre reinforcement, the crack propagation continued gradually, caused by the gradual damaging of the interlocking along the open-crack length. In these tests, more than 60,000 cycles were necessary for the crack to reach the inter-face. In that sense, once again, the high bond fibres provided the best service. On the other hand, without reinforcement the crack propagated very quickly.

The evolution of the measured crack width versus the debonded length is presen-ted in Figure 6.20. The crack opening for a 160-mm debonded length (80 mm each side of the initial crack) after two millions cycles remained limited. It did not ex-ceed 0.2 mm. In the case of high bond metal fibres it was even less than 0.15 mm. The crack opening at debonding initiation, about 0.05 mm, reflected the effect of shrinkage pre-cracking.

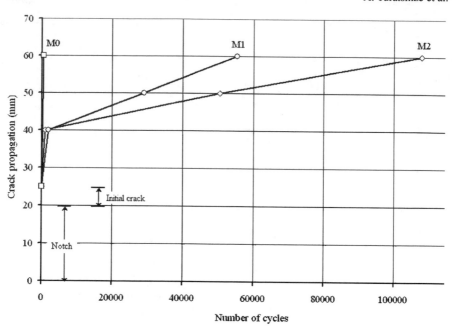

Fig. 6.19 Shrinkage-pre-cracking and fatigue tests: crack propagation through the depth of the overlay versus the number of cycles [21, 55, 57]

All these results converge to indicate that the prevention of debonding initiation or propagation must be achieved as soon as a crack opens and certainly before a 0.2-mm crack opening. The European Standard 1504-3 "Products and systems for the protection and repair of concrete structures – Structural bonding" also confirms this point. It systematically accompanies the limit bond strength requirements by the statement "... with maximum average crack width ≤ 0.05 mm with no crack ≥ 0.1 mm".

6.9 Special Overlays

6.9.1 Stang and Walter's Results [11, 64, 65]

Stang and Walter developed and investigated a cement-based overlay stiffening a steel structure, typically an orthotropic steel bridge deck. Their studies included experiments and modeling that emphasized similarities and differences with more conventional overlay systems.

The specimens were made of a steel plate, 8 mm thick, overlaid with a bonded layer, 40 mm thick, of self-compacting fibre reinforced cement-based material. The bond was ensured by pouring the self-compacting cement-based material onto the

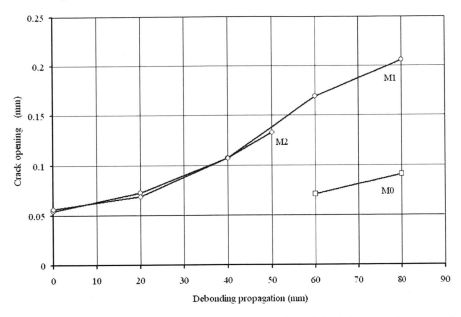

Fig. 6.20 Crack opening versus debonded length each side of the initiating point: case of shrinkage-pre-cracking plus fatigue loading. M0: plain mortar, M1: reinforced with 40 kg/m^3 of slipping fibres and M2: reinforced with 20 kg/m^3 of high bond fibres [21, 55, 57]

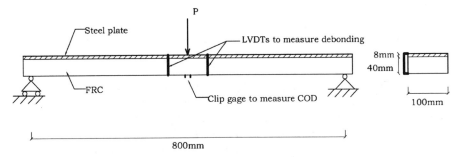

Fig. 6.21 Schematically experimental set-up simulating a cement-based overlays (in this case FRC) cast on a steel plate [65]

sand blasted steel plate. Since crack formation in the overlay and associated debonding under negative bending was of particular concern, a test set-up was used, simulating a part of a stiffening overlay cast on a steel bridge deck, loaded in negative bending, using a composite beam turned up-side down (Figure 6.21).

Two main classes of fibre reinforced materials were tested: strain softening and hardening. Furthermore, in the strain softening class, two materials were tested, ordinary Fibre Reinforced Concrete (FRC) and Fibre Reinforced DENSITÏ (FRD). The two strain softening materials differ in their different stress-crack opening relationships (Figure 6.22a) determined in uniaxial tension using notched specimens

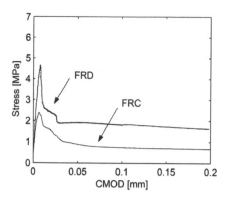 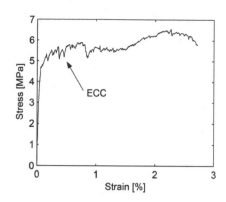

Fig. 6.22 (a) Uniaxial response for Fibre reinforced DENSITÏ (FRD) and Fibre reinforced concrete (FRC); (b) Uniaxial response for ECC [65]

according to the RILEM standard. The FRD material is stronger than FRC and supports, experimentally, the hypothesis that debonding is closely related to formation of bending cracks in the overlay. Furthermore, a special strain hardening material, an Engineered Cementitious Composite (ECC), was tested [66]. Figure 6.22b shows its response in uniaxial tension. Note that due to the strain hardening characteristics of the material and the distributed nature of the crack pattern, the uniaxial response in tension is presented as stress vs. strain. The motivation for using this material was to determine how debonding is affected in the case where no macro bending cracks are formed in the overlay. As for this material, when exposed to uniaxial tension and also in negative bending, multiple cracking was formed in a strain range from 0–3%.

The results from the bending tests with composite beams are presented in a load versus deflection diagram (Figure 6.23). A second y-axis couples debonding of the overlay to the overall response. The study supports the fact that debonding is initiated from a defect or macro crack in the overlay. Furthermore it was shown that when no macro-crack formed in the overlay, which is the ECC case, until a deflection of about 15 mm debonding was not initiated. In the case of FRD, debonding was initiated for a higher load level due to the higher tensile strength and resistance to crack propagation; however debonding was initiated for the same amount of deflection.

All cases have been modelled numerically using finite element software [65]. The studies concluded that the interfacial interlocking effect, mixed normal and shear stresses, has to be taken into account.

The proportions of the investigated specimens can explain the difference with the other results, especially results not reported here related to crack opening versus debonding initiation and propagation. The overlay was several times thicker than the substrate as opposed to conventional overlay systems. Consequently, the overlay-substrate interface was on the compression side of the composite specimen. Initially, the whole specimen behaved as a fibre-reinforced concrete beam with a steel skin on

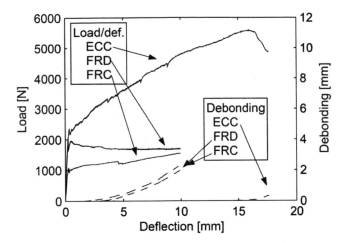

Fig. 6.23 Experimental results for three fibre reinforced cement-based materials [65]

its compression side. Debonding was expected "late" when a macro crack propag-ated deeply enough through the overlay, initiating tensile stresses perpendicular to the interface.

6.9.2 Anchoring of the Overlay

In the continuation of the tests by Gagné et al. [61], Guindon [63] and by Lemieux et al. [62] presented in Section 6.7.2, Benzerzour [67] used the same experimental approach as the one used by Guindon [63] and Lemieux et al. [62] to analyze the effect of mechanical anchorages on the performance of overlays located in a ten-sion zone. Anchors were commercial steel rods having a length of 114 mm and a diameter of 9.5 mm. They were fixed to the substrate by inserting them into drilled holes (84 mm depth) partially filled with a special epoxy resin.

A reference reinforced concrete slab (3 × 3 × 0.18 m) was repaired with a 40-mm bonded overlay (non-anchored) made with a 48 MPa air-entrained concrete. Two other slabs were overlaid with the same type of concrete (thickness of 40 mm) but with different anchor configurations. One configuration consisted of a uniform pattern of anchors equally spaced every 0.2 m in the longitudinal direction and every 0.13 m in the transverse direction. The other configuration was similar except that the longitudinal spacing was reduced to 0.1 m in a 1-m long central segment of the slab.

All the slabs were subjected to cycling loading according to the procedure used by Guindon [63] and Lemieux [62]. For the reference slab, fatigue failure occurred after 490,000 loading cycles. During the test, several debonding cracks appeared on the lateral faces of the slab in the midspan section [67]. No sign of debonding was found for either type of anchored overlays. The flexural cracking patterns were the same for the three slabs, indicating that no preferential cracking was induced by the anchor pattern. These initial results suggested that the use of mechanical anchors is technically possible and could help reduce the risk of debonding. However, more research is needed to optimize the anchor pattern and to assess the cost/benefit ratio of this approach more precisely.

A Swedish experience of overlay anchoring is also discussed in Section 4.4.8.

6.10 Boundaries and Joints of the Overlays

6.10.1 Boundaries and Full Depth Joints

Boundaries and full depth joints are the most critical zones because lack of reinforcement to compensate for the associated discontinuity. Consequently, they must be placed at locations of lowest stressing.

Debonding is the consequence of built-in stresses coming from the cumulative effects of length changes, essentially shrinkage, and negative moment putting the overlay in tension.

To have the least stress at the boundaries or joints, they must be located where there is no negative moment, i.e. exactly with the top of the joints of the substrate.

Else, if the overlay is continuous over a joint of the substrate, there is a great risk that the unavoidable moves of the substrate's joint induces a reflected crack in the overlay. Consequently, as far as possible the edge and joints pattern of the overlay should reflect the one of the substrate.

The through cracks of the substrate, if not carefully reinforced, will behave as joint.

6.10.2 Sawn Joints

If, as is recommended, the overlay is fibre-reinforced, the depth of the joint must be the least possible in order to preserve the maximum force transmission capacity through this weakened section. A reasonable depth is at least a quarter and no more than a third of the overlay thickness.

If necessary such fibre reinforced joints do not have to always match joints of the substrate. If the overlay is reinforced by steel bars or welded fabric, the saw cut must preserve a sufficient cover of the steel.

6.11 Conclusion

Debonding initiates from a mechanical discontinuity (e.g. a crack, a joint or an edge) in an overlay tensioned by the combined effects of the mechanical loading of the structure and of the differential length changes between the overlay and its substrate. From the discontinuity, debonding is always initiated by tension perpendicular to the interface. The main origin of the debonding tensile stress perpendicular to the interface is the peeling effect. Secondarily, the geometrical effect resulting from the curvature of the structure placing the overlay in tension (Figure 6.1) and the curling trend of the overlay because of its non uniform shrinkage add to the effects. Consequently, the relevant characteristic of the interface to prevent debonding is its tensile strength perpendicular to the interface.

Reinforcement of the overlay by fibres, steel bars, or welded fabric sustains the ability to maintain a partial continuity of through-crack force transmission. This feature acts positively to delay or prevent debonding.

When modeling the debonding mechanism, an essential feature is to take into account the interlocking effect, which exists between the two faces of the crack and of the debonded interface.

Tests or modeling in "first monotonic loading or straining" condition are efficient to investigate the debonding mechanism. However they fail to predict behaviour and durability in-situ conditions. In these last conditions only "shrinkage-pre-cracking and fatigue" investigations are relevant. The reason is that, on the one hand, pre-cracking debonding may occur at lower curvature level. On the other hand, fatigue significantly alters interlocking.

The beneficial effect of fibre reinforcement is almost inconsequential in the case of first monotonic loading. On the other hand, it has a major effect in the case of "shrinkage-pre-cracking and fatigue". Moreover, debonding has to be prevented in the early phase of cracking. When the crack width reaches 0.2 mm, it is too late to prevent debonding. Fibre reinforced cement-based overlays are highly recommended, and high bond fibres provide the best performance.

References

1. Carter, A., Gurjar, S. and Wong, J., Debonding of highway bridge deck overlays, *Concrete International*, **24**(7), 51–58, 2002.
2. Silfwerbrand, J., Improving concrete bond in repaired bridge decks, *Concrete International*, **12**(9), 61–66, 1990.
3. Silfwerbrand, J. and Paulsson, J., Swedish experience: Better bonding of bridge deck overlays, *Concrete International*, **20**(10), 56–61, 1998.
4. Saucier, F. and Pigeon, M., Durability of new-to old concrete bondings, in *Proceedings of ACI International Conference on Evaluation and Rehabilitation of Concrete Structures and Innovations in Design (ACI SP-128)*, Hong Kong, December, pp. 689–705, 1991.
5. Langlois, M., Pigeon, M., Bissonnette B. and Allard D., Durability of pavement repairs: A field experiment, *Concrete International*, **16**(8), 39–43, 1994.

6. Delatte, N.J., Wade, D.M. and Fowler, D.W., Laboratory and field testing of concrete bond development for expedited bonded concrete overlays, *ACI Materials Journal*, **97**(3), 272–280, 2000.
7. Tschegg, K.E., Igruber, M., Srberg, C.H. and Münger, F., Factors influencing fracture behavior of old-new concrete bonds, *ACI Materials Journal*, **97**(4), 447–453, 2000.
8. Vaysburd, A.M. and Emmons, P.H., How to make today's repairs durable for tomorrow-corrosion protection in concrete repair, *Construction and Building Materials*, **14**, 189–197, 2000.
9. Emmons, P.H., Vaysburd, A.M. and Czarnecki, L., Durability of repair materials: Current practice and challenges, in *Brittle Matrix Composites 6, Proceedings of an International Symposium*, Warsaw, pp. 263–274, 2000.
10. Sabathier, V., Granju, J-L., Turatsinze, A. and Bissonnette B., Repair by cement-based thin overlays – Interlocking at the interface and modeling of debonding, in *Industrial Floors'03, Proceedings of an International Colloquium*, Esslingen, January, P. Seidler (Ed.), pp. 621–626, 2003.
11. Walter, R., Stang, H., Olesen, J.F. and Gimsing, N.J., Debonding of FRC composite deck bridge, in *Brittle Matrix Composites 7, Proceedings of an International Symposium*, Warsaw, October, Woodhead Publishing, pp. 191–200, 2003.
12. Lupien, C., Chanvillard, G. Aïtcin, P-C. and Gagné, R., Réhabilitation d'une chaussée par resurfaçage mince adhérent en béton renforcé de fibres d'acier, in *Proceedings of AIPCR, Comité C-7*, Montréal, Canada, pp. 246–250, 1995.
13. Scott, M., Rezaizadeh, A., Delahaza, A., Santos, C.G., Moore, M., Graybeal, B. and Washer G., A comparison of non destructive evaluation methods for bridge deck assessment, *NDT & E International*, **36**(4), 245–255, 2003.
14. Clark, M.R., McCann, D.M. and Forde, M.C., Application of infrared thermography to the non-destructive testing of concrete and masonry bridges, *NDT & E International*, **36**(4), 265–275, 2003.
15. Lupien, C., Chanvillard, G. Aïtcin, P-C. and Gagné R., Réhabilitation d'une chaussée en béton avec une chape mince en béton renforcé de fibres d'acier, in *Les techniques de transport au service de la qualité de vie,Exposé des communications du 25° congrès annuel de l'AQTR*, Montréal, April, pp. 108-122, 1990.
16. Bungey, J.H., Sub-surface radar testing of concrete: A review, *Construction and Building Materials*, **18**, 1–8, 2004.
17. Dalhuisen, D.H., Stroeven, P., Moczko, A.T. and Peng, X., Modern testing methods for "insitu" non-destructive examination of concrete structures, in *Proceedings of The NOCMAT/3-Vietnam 3rd International Conference of Non-Conventional Materials and Technologies*, Hanoi, Vietnam, March, K. Ghavami and Nguyen Tien Dich (Eds.), pp. 253-259, 2002.
18. Moczko, A.T., Pszonka, A. and Stroeven, P., Acoustic emission as a useful tool for reflecting cracking behaviour of concrete composites, in *Proceedings of International Symposium Non-Destructive Testing in Civil Engineering (NDT-CE)*, Berlin, September, G. Schickert and H. Wiggenhauser (Eds.), pp. 805–812, 1995.
19. Chausson, H. and Granju J.-L., Optimized design of fiber reinforcement thin bonded overlays, in *Brittle Matrix Composites 5, Proceedings of an International Symposium*, Warsaw, October, Woodhead Publishing, pp. 133–142, 1997.
20. Chowdhury, M.R. and Ray, J.C., Accelerometers for bridge load testing, *NDT & E International*, **36**(4), 237–244, 2003.
21. Farhat, H., Durabilité des réparations en béton de fibres: Effets du retrait et de la fatigue, PhD Thesis, Université Paul Sabatier, Toulouse, France, 178 pp., 1999 [in French].
22. Basu, B., Identification of stiffness degradation in structures using wavelet analysis, *Construction and Building Material*, **19**(9), 713–721, 2005.
23. Granju, J-L., Debonding of thin cement-based overlays, *Journal of Materials in Civil Engineering*, **13**(2), 114–120, 2001.
24. Bernard, O., Comportement à long terme des éléments de structure formés de bétons d'âges différents, PhD Thesis No. 2283, Ecole Polythechnique Fédérale de Lausanne, Switzerland, 189 pp., 2000.

25. Bigwood, D.A. and Grocombe, A.D., Elastic analysis and engineering design formulae for bonded joints, *International Journal of Adhesion and Adhesive*, **9**, 229–242, 1989.
26. Fowler, D.W., Wheat, D.L., Choi, D.U. and Zalatimo, J., Stresses in PC overlays due to thermal changes, in *Industrial Floors'03, Proceedings of an International colloquium*, Esslingen, January, P. Seidler (Ed.), pp. 29–36, 2003.
27. Naciri, T., Ehrlacher, A. and Chabot, A., Interlaminar stress analysis with a new multiparticle modelisation of multilayered materials (M4), *Composites Sciences and Technology*, **58**(3), 337–343, 1998.
28. Caron, J.F., Diaz Diaz, A., Carreira, R.P., Chabot, A. and Ehrlacher, A., Multi-particle modelling for prediction of delamination in multi-layered materials, *Composites Sciences and Technology*, **66**(6), 755–765, 2006.
29. Granju, J.-L., Thin bonded overlays: About the role of fiber reinforcement on the limitation of their debonding, *Advanced Cement Based Materials*, **4**(1), 21–27, 1997.
30. Balouch, S.U. and Granju, J.-L., Corrosion of different types of steel fibres in SFRC and testing of corrosion inhibitors, in *Infrastructure Regeneration and Rehabilitation – Improving the Quality of Life through Better Construction. A Vision for the Next Millennium, Proceedings of an International Conference*, Sheffield, June–July, Sheffield Academic Press, pp. 735–747, 1999.
31. Silfwerbrand, J., Shear Bond Strength in repaired concrete structures, *Materials and Structures*, **36**(260), 419–424, 2003.
32. Sabathier, V., Rechargements minces adhérents à base cimentaire renforcés de fibres métalliques. Conditions de leur durabilité, modélisation et calcul, PhD Thesis, Université Toulouse III, 190 pp., 2004.
33. Julio, E.N.B.S, Branco, F.A.B. and Silva, V.D., Concrete-to-concrete bond strength. Influence of the roughness of substrate surface, *Construction and Building Materials*, **18**(9), 675–681, 2004.
34. Turatsinze, A., Farhat, H. and Granju, J-L., Durability of metal-fibre reinforced concrete repairs: Drying shrinkage effects, in *Proceedings of an International Symposium*, Warsaw, October, Woodhead Publishing , pp. 296–305, 2000.
35. Granju, J.L., Sabathier, V., Turatsinze, A. and Toumi, A., Interface between an old concrete and a bonded overlay: Debonding mechanism, *Interface Science Journal*, **12**(4), 381–388, 2004.
36. Grzybowski, M. and Shah, P.S., Shrinkage cracking of fiber reinforced concrete, *ACI Materials Journal*, **87**(2), 138–148, 1990.
37. Marosszeky, M., Stress performance in concrete repairs, in *Proceedings of a RILEM International Conference on Rehabilitation of Concrete Structures*, Melbourne, pp. 467–474, 1992.
38. Saucier, F. and Pigeon, M., Testing of superficial repairs for sidewalks in Canada, *Concrete International*, **18**(5), 39–43, 1996.
39. Banthia, N., Yan, C. and Mindess, S., Restrained shrinkage cracking in fiber reinforced concrete: A novel test technique, *Cement and Concrete Research*, **26**(1), 9–14, 1996.
40. Mailvaganam, N., Springfield, J., Repette, W. and Taylor, D., Curling of concrete slabs on grade, *Construction Technology Update*, **44**, 1–6, 2000.
41. Suprenant, B.A. and Malisch, R.W., Repairing curled slabs, *Concrete Construction*, **9**, 58–65, 1999.
42. Suprenant, B.A., A look at the curling mechanism and the effect of moisture and shrinkage gradients on the amount of curling, *Concrete International*, **24**(3), 56–61, 2002.
43. Sabathier, V., Granju, J-L., Bissonnette, B., Turatsinze, A. and Tamtsia, B., Cement-based thin bonded overlays: Numerical study of the influence of a bond defect, in *Brittle Matrix Composites 7, Proceedings of an International Symposium*, Warsaw, October, Woodhead Publishing, pp. 181-189, 2003.
44. Tran, Q.T., Toumi, A. and Turatsinze, A., Durability of an overlay-old concrete interface: The role of a metal fibre reinforcement, in *Brittle Matix Composites 8, Proceedings of an International Symposium*, Warsaw, October, Woodhead Publishing, pp. 409–419, 2006.

45. Tran, Q.T., Toumi, A. and Turatsinze, A., Thin bonded cement-based overlays: Numerical analysis of factors influencing their debonding under fatigue loading, *Materials and Structures*, **41**(5), 951–967, 2008.

46. Betterton, R.M., Knutson, M.J. and Marks, V.J., Fibrous portland cement concrete averlay research in Green County, Iowa, Transportation Research Record, No. 1040, TRB, National Research Council, Washington DC, pp. 1–7, 1985.

47. Paulsson, J. and Silfwerbrand, J., Durability of repaired bridge deck overlays, *Concrete International*, **20**(2), 76–82, 1998.

48. Verhoeven, K., Thin overlays of steel fiber reinforced concrete and continuously reinforced concrete, state of the art in Belgium, in *Proceedings of the 4th International Conference on Concrete Pavement Design and Rehabilitation*, Purdue University, West Lafayette, IN, April, pp. 205–219, 1989.

49. Chanvillard, G., Aitcin, P.C. and Lupien, C., Field evaluation of steel-fibre reinforced concrete overlays with bonding mechanism, in *Transportation Research Record 1226, TRB, Washington*, pp. 48–56, 1990.

50. Chanvillard, G. and Aitcin, P.C., Thin bonded overlays of fiber-reinforced concrete as a method of rehabilitation of concrete roads, *Canadian Journal of Civil Engineering*, **17**(4), 521–527, 1990.

51. Belaghmas, A., Fissuration et décollement d'une couche de béton adhérente à un support, DEA memory, LMDC, Génie Civil INSA-UPS, Toulouse, 1993.

52. Granju, J.-L. and Chausson, H., Serviceability of fiber reinforced thin overlays relation between cracking and debonding, in *ConChem, Proceedings of an International exhibition & Conference*, Brussels, November, Verlag für chemische industrie, pp. 133–142, 1995.

53. Granju, J.-L. and Chausson, H., Fiber reinforced thin bonded overlays: The mechanism of their debonding in relation with their cracking, in Concrete repair, rehabilitation and protection, in *Proceedings of an International Congress*, Dundee, June, E. & FN. Spon, pp. 583–590, 1996.

54. Chausson, H., Durabilité des rechargements minces en béton: Relation entre leur décollement, leur fissuration et leur renforcement par des fibres, PhD Thesis, Université Paul Sabatier, Toulouse, France, 198 pp., 1997 [in French].

55. Turatsinze, A., Granju, J.L., Sabathier, V. and Farhat, H., Durability of bonded cement-based overlays: effect of metal fibre reinforcement, *Materials and Structures*, **38**(277), 321–327, 2005.

56. Habel, K., Brühwiler, E. and Bernard, O., The numerical investigation of delamination in hybrid reinforced concrete elements, in *Proceedings of International PhD Symposium in Civil Engineering*, Vienna, October, K. Bergmeister (Ed.), pp. 221–228, 2000.

57. Farhat, H., Turatsinze, A. and Granju, J.-L., Durabilité des rechargements minces adhérents soumis à la fatigue mécanique, in *Proceedings (CD-rom) of 14ième Congrès Français de Mécanique*, Toulouse, August–September, 1999.

58. Turatsinze, A., Farhat, H. and Granju, J.-L., Influence of autogenous cracking on the durability of repairs by cement-based overlays reinforced with metal fibres, *Materials and Structures*, **36**(264), 2003, 673–677.

59. Zhang, J., Stang, H. and Li, V.C., Crack bridging model for fibre reinforced concrete under fatigue tension, *International Journal of Fatigue*, **23**(8), 655–670, 2001.

60. Rossi, P. (sous la direction de), *Le développement industriel des bétons de fibres métalliques, conclusions et reconclusions*, Presse de l'école nationale des Ponts et Chaussées, 2002.

61. Gagné, R., Bissonnette, B., Lachemi, M. and Lemieux, M., Vézina, Analyse du comportement de resurfaçages adhérents utilisés pour réparer des dalles en béton armé, in *9e Colloque sur la progression de la recherche québécoise sur les ouvrages d'art*, Québec, Mai, 11 pp.

62. Lemieux, M., Gagné, R., Bissonnette, B. and Lachemi, M., Behavior of overlaid reinforced-concrete slab panels under cyclic loading – Effect of interface location and overlay thickness, *ACI Structural Journal*, **102**(3), 454–461, 2005.

63. Guindon, M.-A., Étude du comportement des resurfaçages adhérents – Mécanismes d'endommagement et influence des paramètres de conception, Mémoire de maîtrise, Université de Sherbrooke, Département de génie civil.

64. Walter, R., Stang., H, Gimsing, N. J. and Olesen, J.F., High performance composite decks using SCSFRC, in *Proceedings Fourth International Workshop on High Performance Fiber Reinforced Cementitious Composites*, Ann Arbor, June, pp. 495–504, 2003.
65. Walter, R., Li, V.C. and Stang, H., Comparison of FRC and ECC in a composite bridge deck, in *5th International PhD Symposium in Civil Engineering*, Delft, the Netherlands, June, pp. 477–484, 2004.
66. Li, V.C., *Advances in ECC Research*, ACI Special Publication on Concrete: Material Science to Applications, SP 206-23, pp. 373–400.
67. Benzerzour, M., Étude expérimentale et numérique du renforcement des tabliers de pont en béton armé par des resurfaçages adhérents, PhD Thesis, Université de Sherbrooke, Département de génie civil, Canada [in French].

Chapter 7
Design

M. Treviño, J.-L. Granju, H. Beushausen, A. Chabot, H. Mihashi
and J. Silfwerbrand

Abstract This chapter reviews the design methods typically use of designing Bonded Concrete Overlays (BCO). Two levels of design are discussed: the design for strength and the design to resist debonding. For the design for strength various methods are discussed: U.S. Army Corps of Engineers, Portland Cement Association, American Association of Highway and Transportation Officials, reinforcement design of overlaid continuously reinforced concrete pavements and procedures from other countries. For design to resist debonding, methods from the U.S., Sweden, Japan, other countries in Europe are described.

7.1 Introduction

The objective of this chapter is to review the design methods, codes or recommendations currently in use for Bonded Concrete Overlays (BCO) for repairs, such as pavement rehabilitation.

M. Treviño
Center for Transportation Research, The University of Texas at Austin (TX), U.S.A.

J.-L. Granju
Laboratoire Matériaux et Durabilité des Constructions (LMDC), UPS-INSA, Toulouse, France

H. Beushausen
Department of Civil Engineering, University of Cape Town, Cape Town, South Africa

A. Chabot
Laboratoire Central des Ponts et Chaussées, Nantes, France

H. Mihashi
Department of Architecture and Building Science, School of Engineering, Tohoku University, Sendai, Japan

J. Silfwerbrand
Swedish Cement and Concrete Research Institute (CBI), SE-100 44 Stockholm, Sweden

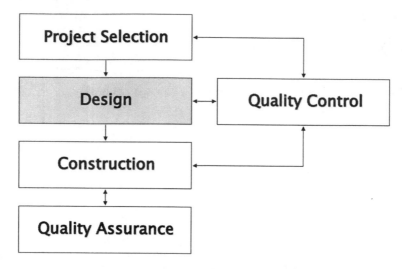

Fig. 7.1 Design stage as part of the BCO process

Two levels of design must be distinguished.

- The *design for sufficient strength* of the overlaid structure. For BCO, it assumes a perfect bond between the overlay and the substrate. If that is the case, the design procedures are similar to those of monolithic structures. Also, depending on the code, they may include the evaluation of the structure to be overlaid and of its contribution to the strength in the final overlaid structure.
- The *design to prevent debonding*. Its role is to ensure a monolithic behaviour of the overlaid structure (required in the first level above) and its durability. The calculations are applicable to composite structures, and the behaviour of the overlay-substrate interface is a prevalent parameter.

7.2 Design for Sufficient Strength

If a BCO project is represented as a series of stages, as illustrated in Figure 7.1, the design would correspond to the second phase of the BCO process, following the project selection and preceding the construction. Once a BCO has proven an appropriate rehabilitation choice for the pavement in question, during the project selection stage, the next step is to propose a thickness design for the overlay and a reinforcement design. The thickness design is based upon the condition of the existing pavement, the purpose of the overlay, the projected design life for the rehabilitated structure, the historic and projected traffic data, and the material properties of the existing pavement, as well as those of the new overlay.

The following discussion is focused on the common pavement BCO design methodology utilized in the United States. However, the procedures described herein are also applicable to other cases, such as bridge decks or slabs on grade.

Among the methods presented, the AASHTO procedure distinguishes itself by the implementation of an interesting concept, the "Condition Factor" (CF) which accounts for an eventual weakening (due to fatigue) of the overlaid substrate.

This section is comprised of two subsections. The first covers a discussion of some of the most utilized BCO thickness design methods. The second presents reinforcement design procedure for BCOs on continuously reinforced concrete pavements (CRCP).

7.2.1 Overview

The overlay design is not very different from the design of a new pavement. Many of the concepts involved in the design of a new pavement apply to a BCO design as well.

7.2.1.1 Purpose of the Overlay

The rehabilitation is aimed to remedy a specific situation, which the designer knows from the project selection phase. The need for rehabilitation arises as the pavement experiences a decrease in serviceability; this occurs as a consequence of load applications, age, and traffic increases. The interaction of these factors is manifested as structural and functional failures. When the overlay is being placed for the purpose of structural improvement, the required thickness of the overlay is a function of two components: the structural capacity of the existing pavement under current conditions, and the structural capacity necessary to fulfill future traffic demands.

However, if the pavement deterioration is a consequence of non-loading factors, i.e., the purpose of the overlay is only functional improvement, the design equations will render a minimal or zero thickness, in which case, a minimum constructible BCO thickness will be enough to address the functional deficiency. BCOs as thin as one inch ($\approx$ 25 mm) have been used successfully on sound pavements [1].

7.2.1.2 Basic Design Principle

The foremost structural characteristic that differentiates a BCO from other rehabilitation concepts is that, by definition, the overlay behaves as a single unit in conjunction with the existing pavement. Therefore, the structural capacity remaining in the existing substrate is fully utilized. As such, it is accounted for in the design equations, which contributes to reduce the thickness of the overlay required. This is only attainable if the bond between overlay and substrate is achieved and maintained.

Therefore, if the purpose of the BCO is to remedy structural deficiencies, the design is based on a simple equation: the BCO is designed by determining the additional thickness of concrete needed to carry the anticipated traffic. Thus, the equation is as follows:

$$D_{\text{BCO}} = D_f - D \qquad (7.1)$$

where D_{BCO} is the overlay thickness, D_f is the required thickness to carry the future traffic if the pavement were constructed new, and D is the effective existing pavement slab thickness. The design methods vary in the way in which the existing pavement contribution, D_f, is determined and the way the original thickness is affected by the pavement condition.

7.2.2 Design Concepts

A great deal of information and decisions necessary for new construction projects are also required in rehabilitation projects such as BCOs. Many of the following concepts apply to both new construction and rehabilitation projects.

7.2.2.1 Overall Standard Deviation

Since there is variability inherent in traffic predictions, material properties, quality control, and environmental conditions during construction, it is reasonable to use a probabilistic approach in the design, rather than a deterministic one. The overall standard deviation accounts for all the sources of uncertainty involved in the overlay design, like the material properties and traffic data, adding flexibility to the design. If the information corresponding to the materials characterization and traffic is deemed accurate, the data will have a small standard deviation, rendering a thinner overlay. On the other hand, if the information were not collected properly, or is not entirely available or trustworthy, the data will have a large standard deviation resulting in a thicker overlay. The AASHTO Guide [2] recommends a standard deviation of 0.39 for PCC overlays.

7.2.3 Current Overlay Design Procedures

There are several BCO design procedures available. The basic concepts of some of the most common procedures are discussed below.

7.2.3.1 Corps of Engineers

The U.S. Corps of Engineers procedure was originally devised for the design of PCC overlays over PCC airfield runways and taxiways. It was developed using full-scale accelerated test tracks [3].

This method uses a version of the basic design equation (7.1), modified by empirical coefficients. In this method, the required thickness of the overlay is the difference between the thickness required for a new pavement and the thickness of the existing slab. In Equation (7.1), instead of the thickness of the existing slab, the thickness considered is the effective thickness. Three models were developed, namely, for the bonded, partially bonded and unbonded cases, represented by Equations (7.2), (7.3) and (7.4), respectively. Even though the scope of this chapter covers only the bonded overlay case, the other two are presented as a reference to provide a better understanding of the methodology.

$$h_o = h_n - h_e \tag{7.2}$$

$$h_o = \sqrt[1.4]{h_n^{1.4} - Ch_e^{1.4}} \tag{7.3}$$

$$h_o = \sqrt{h_n^2 - Ch_e^2} \tag{7.4}$$

where h_o is the required overlay thickness; h_n is the required theoretical thickness for the design loading, for a new pavement; h_e is the existing pavement thickness; and C is the condition correction coefficient.

The values for C are determined by visual inspection, and range between 1 (for a pavement in good structural condition) to 0.35 (for a pavement with severe structural defects).

As can be inferred by these equations, it is assumed that, for a BCO, the condition correction coefficient equals one, which means that the existing slab is in good structural condition or will be upgraded to this condition. Besides this coefficient, the other difference among these equations is the exponent of the thicknesses, which is related to the bonding characteristics of each type of overlay. For a BCO, the value of the exponent is 1, given that the BCO and the substrate will behave monolithically. In a similar fashion, it is equal to 1.4 and 2 in Equations (7.3) and (7.4), for the partially bonded and unbonded overlays, respectively.

In this method, it is implied that the existing concrete has suffered no fatigue due to traffic or other factors, and it is as strong as the concrete in a new pavement, which contradicts the fatigue damage concept explained above and the idea of remaining life. Furthermore, it assumes that the failure mechanism of the overlaid pavement is the same as that of a new pavement.

Failure is defined as the load application level at which cracking or structural breakup first appears, which does not apply properly to highway concrete overlays, where cracking is an inherent occurrence of portland cement concrete pavements.

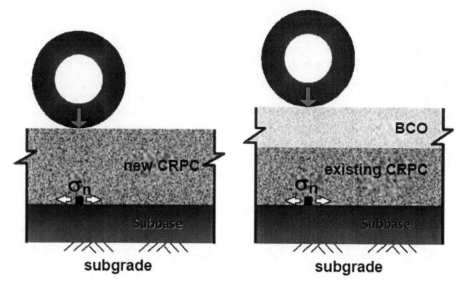

Fig. 7.2 Edge stresses for new and overlaid pavement for PCA method design equivalency

7.2.3.2 Portland Cement Association

The PCA methodology consists in designing an overlay system that is structurally equivalent to a new full-depth pavement placed on the same subbase and subgrade. Unlike the Corps of Engineers procedure, it uses an evaluation of the existing pavement by means of condition surveys, deflection tests, and in-situ sample testing, to take its condition into consideration in the design.

The design basis is the analysis of the stresses at the edge of the pavement [4], i.e., the stress at the bottom of the slab along the lateral edge of the slab plate. The model equates the edge stress at the bottom of the new full-depth pavement (σ_n) with that of the overlaid system at the bottom of the existing pavement (σ_e), as shown in Figure 7.2.

Because the new full-depth slab and the existing concrete will have different moduli of rupture, S_c, the equivalency is based on the stress ratio to the modulus of rupture. If the stress ratio for the overlaid system is the same than that of the new pavement, both pavements will be structurally equivalent:

$$\frac{\sigma_n}{S_{cn}} \geq \frac{\sigma_e}{S_{ce}} \tag{7.5}$$

where σ_n is the critical stress edge in the new pavement; S_{cn} is the modulus of rupture of the new concrete; σ_e is the critical stress edge in the existing pavement; and S_{ce} is the modulus of rupture of the existing concrete.

In developing this method, a finite element program was used to create a design chart in which the critical tensile stresses due to edge loading in both new pavement and the BCO structure are related to the modulus of rupture of the existing concrete, for which three different ranges of moduli are considered.

For the BCO design, the first step consists of calculating the thickness of the new full-depth pavement, and this can be accomplished by using the PCA design method or other PCC design method. With this thickness and the design chart, the combined thickness of BCO and existing pavement is computed, and the BCO thickness is determined by subtracting the existing slab thickness from this value. The maximum BCO thickness recommended is 5 in ($\approx$ 125 mm). When the required thickness exceeds this value the use of an unbonded overlay is more advisable.

In this method, the fatigue consideration is dependent on the procedure used to arrive at the new full-depth pavement thickness. If the Portland Cement Association method is used, then it is assumed that the pavement can withstand an infinite number of applications, as long as those occur under the stress limit established by the method, which is based on plain concrete tests.

7.2.3.3 AASHTO

The AASHTO method, outlined in detail in the 1993 AASHTO Guide [2], is based upon the AASHO Road Test, the basic design equation (7.1), the remaining life concept, and thus, fatigue, and the idea of serviceability.

It is mostly an empirical method, since the design equations for the method were derived from regression analyses performed on the Road Test data, but it includes a mechanistic part, in the determination of stresses and strains. Like the PCA method, the AASHTO procedure advocates conducting a comprehensive evaluation of the existing pavement conditions, and applying the results as input design parameters for the BCO.

The thickness design equation is:

$$D_{ol} = D_f - D_{eff} \tag{7.6}$$

where D_{ol} is the required thickness of BCO; D_f is the slab thickness to carry future traffic; D_{eff} is the effective thickness of existing slab, calculated by applying a condition factor (CF) to the existing slab thickness, D, as in the following expression:

$$D_{eff} = CF \times D \tag{7.7}$$

The value of CF can be determined in two ways, either by the use of remaining life or by means of the condition survey.

The value of CF ranges from 0 to 1. When the pavement condition is satisfactory, it will take the value of one, or close to one, which means that the condition of the pavement does not affect the effective thickness. However, as the condition of the slab is more deteriorated, the value of the condition factor decreases. Guidelines for selecting values for CF appear in AASHTO 1993.

The slab thickness to carry future traffic, D_f, is calculated as if it were a new pavement design.

As a reference to the reader, the AASHTO design equation for rigid pavements is provided below:

$$\log_{10} W_{18} = Z_R \times S_O + 7.35 \times \log_{10}(D+1) - 0.06 + \frac{\log_{10}\left[\frac{\Delta\,PSI}{4.5-1.5}\right]}{1 + \frac{1.624 \times 10^7}{(D+1)^{8.46}}}$$

$$+ (4.22 - 0.32 p_t) \times \log_{10}\left[\frac{S'_c \times C_d \times (D^{0.75} - 1.132)}{215.63 \times J\left[D^{0.75} - \frac{18.42}{\left(\frac{E_c}{k}\right)^{0.25}}\right]}\right]$$

$$(7.8)$$

where W_{18} is the predicted number of 18-kip ESAL applications; Z_R is the standard normal deviate; S_O is the overall standard deviation of rigid pavement; D is the thickness of pavement slab (in inches); $\Delta\,PSI$ is the difference between initial serviceability, p_o, and terminal serviceability index, p_t; S'_c is the PCC modulus of rupture, psi; J is the load transfer coefficient; C_d is the drainage coefficient; E_c is the PCC modulus of elasticity, psi; and k is the modulus of subgrade reaction, pci.

The first term ($Z_R \times S_O$) corresponds to the reliability. The remaining terms on the first line of the equation are the empirical part of the procedure, derived from the data gathered at the AASHO Road Test, while the second line, related to stress computations, is the theoretical part, which was added to account for changes in strength and stresses owing to physical constants (e.g., E_c, k) occurring in conditions other than those that existed during the road test.

7.2.4 Reinforcement Design of Overlaid Continuously Reinforced Concrete Pavements (CRCP)

This section presents the design of the overlay reinforcement. The purpose of the steel reinforcement in a CRCP is to hold cracked concrete together and to maintain load transfer; it does not increase the pavement structural capacity. It is recommended that the BCO be reinforced in a similar fashion as the existing pavement, i.e., similar steel percentage, bar sizes and spacings, unless there is a significant deficiency in the original pavement reinforcement design.

There are three main design parameters that have to be satisfied in the design of longitudinal reinforcement for continuously reinforced concrete (CRC) pavements:

- crack width at freezing temperature;
- maximum steel stress, at the minimum temperature expected to occur during the design life of the pavement; and
- cumulative percentage of transverse cracks spaced below three feet.

The most widespread longitudinal steel design procedure for CRCP was developed by Noble [5]; a computerized system, known as the CRCP computer program incorporated the design method, which was adopted by the AASHTO Guide.

The CRCP program has the capability to predict the time history of crack spacing, crack width, steel and concrete stress for a range of material properties, environmental conditions, and layer thicknesses. In the steel design process, the first step would be to run the program with the existing pavement properties, including the existing steel percentage and past environmental conditions, to try to replicate the existing crack spacing. In the next step, the new pavement structure, with the overlay included, is analyzed with the program, varying the percentage of steel. The output of the program will include the combination of the reflective cracking of the existing pavement plus the new BCO cracking.

The transverse reinforcement can be resolved with the following expression, which relates the percent of transverse steel, P_s, with the slab length, L (ft), the steel working stress, f_s (psi), and the friction factor, F:

$$P_s = \frac{LF}{2f_s} \times 100 \tag{7.9}$$

7.2.5 BCO Design Procedures in Other Countries

The European standard prEN 1504-3 (2001), dedicated to repair systems and works (see Section 7.2.3) ignores the "design for sufficient strength" of BCO.

In France, there is no specific code for BCO design. The design code or recommendations for new slabs or pavements are used as a reference. The design, therefore, assumes a complete bond. There are also some design computer programs especially developed for pavement design that can be used for this purpose. Among those, the program "ALIZE", developed by LCPC (Laboratoire Central des Ponts et Chaussées, France), for multi-layered pavements; implemented for asphalt-based pavements, works also for cement-based pavements.

7.2.6 Summary and Conclusions

In this section, a historical overview of the development of the most widely used design procedures was presented, along with their general equations and input parameters.

The current BCO design procedures have conceptual differences and limitations; therefore, there is not an absolute solution to BCO thickness design. One of the main differences across design methods is the failure criterion. The Corps of Engineers procedure was developed for airports and, as such, it has the limitation that its failure criterion, based on the appearance of structural cracking, is not applicable to highway pavements. It assumes that the existing concrete has had no fatigue damage and will behave just as new concrete. Hence, it does not use any evaluation of the existing pavement. The AASHTO method has its failure criterion based upon serviceability, whereas the Portland Cement Association method is based on a stress limit established for plain concrete.

7.3 Design to Prevent Debonding

7.3.1 Introduction

In Europe, North America and Japan, up to now, strictly speaking, there are no methods or codes that address the overlay design to prevent debonding. North American, European and Japanese standards or recommendations focus on the quality of the overlaying work (subject of Chapters 2, 3, 4 and 8 of this book), expecting that it should result in durable repairs. The "design" aspect is limited to very crude requirements or recommendations of limit values, usually strengths, to achieve or not to exceed. The specific case of reinforced overlays, notably by fibres, is not dealt with.

7.3.2 USA Recommendations

It seems that the only available recommendations related to bonded overlays are from the American Concrete Pavement Association (ACPA) [6]. Although published in 1990, no updates are available yet.

About bond, it states: "Achieving bond is a key to long-term life extension. A (shear) bond strength of 200 pound per square inch (psi) is sufficient to withstand shearing forces and ensure bond is maintained. This number was determined in the laboratory [7]. At bond shear strength of around 200 psi the laminated or bonded beams were nearly as strong as the monolithic ones." Translated into MPa strength, it means that a 1.4 MPa shear strength is expected to be enough to ensure a durable bond. Knowing that in the case of a bond between cement-based materials, the tensile strength of the bond is around half of its shear strength, it can be assumed that according to ACPA, for a durable bonded overlay, it is sufficient to achieve a 0.7 MPa tensile strength of the bond.

Table 7.1 Content of European Standard EN 1504: "Products and systems for the protection and repair of concrete structures"

Reference	Title
1504-1	General scope and definitions
1504-2	Surface protection systems
1504-3	Structural and non structural repair
1504-4	Structural bonding
1504-5	Concrete injection
1504-6	Grouting to anchor reinforcement or to fill external voids
1504-7	Reinforcement corrosion prevention
1504-8	Quality control and evaluation of conformity
1504-9	General principles for the use of products and systems
1504-10	Application of products and systems and quality control of the work

7.3.3 European Requirements

Currently, in Europe, a standard is available for repair works [8]. The ten parts listed in Table 7.1 comprise the standard. Despite the fact that the standard is currently in the developing stages, all ten parts are already available, some as finished European Standards (EN) and others as provisional standards (prEN). In its field of relevance, it rules the national standards.

Several types of overlaying work are considered, which are:

- Type 3.1: concrete restoration by applying mortar by hand;
- Type 3.2: concrete restoration by recasting with concrete;
- Type 3.3: concrete restoration by spraying with concrete or mortar;
- Type 4.4: structural strengthening by adding mortar or concrete;
- Type 7.1: restoring passivity by increasing cover to reinforcement with additional cementitious mortar or concrete; and
- Type 7.2: restoring passivity by replacing contaminated or carbonated concrete.

In the text of the Standard, all the requirements are collected in two tables distinguishing the cases of "structural" and "non-structural" overlaying systems and can be summarized in Table 7.2.

It results that:

- for "structural" applications, achieve a tensile bond strength ≥ 2.0 MPa;
- for "non-structural" applications, achieve a tensile bond strength ≥ 1.5 MPa;
- in all cases the crack width must remain limited; on the test specimens, no crack ≥ 1 mm is accepted.

Note that South Africa follows the British codes, then the European code.

Table 7.2 European Standard prEN 4504-3: "design" requirements for overlaying works

Property	Test method	Requirement		Comments
		Structural	Non-structural	
Compressive strength	EN 12190	30 MPa	20 MPa	
Elastic modulus (only for type 4 overlaying)	prEN13412	20 GPa		
Bond tensile strength	EN 1542 (pull-off test)	2.0 MPa	1.5 MPa	Mean value No single value less than 75% of minimum requirement
Bond tensile strength after freeze-thaw test (if exposure conditions require it)	840-1	2.0 MPa	1.5 MPa	After 50 cycles, with maximum permissible average crack width 0.05 mm with no crack 0.1 mm and no delamination
Bond tensile strength after thunder-shower test (if exposure conditions require it)	840-2	2.0 MPa	1.5 MPa	After 30 cycles, with maximum permissible average crack width 0.05 mm with no crack 0.1 mm and no delamination
Bond tensile strength after dry thermal cycling test (if exposure conditions require it)	840-4	2.0 MPa	1.5 MPa	After 30 cycles, with maximum permissible average crack width 0.05 mm with no crack 0.1 mm and no delamination
Bond tensile strength after restrained shrinkage/swelling test (only for overlaying types 3.1, 3.2 and 4.4)	816-4	2.0 MPa	1.5 MPa	After test, with maximum permissible average crack width 0.05 mm with no crack 0.1 mm and no delamination

7.3.4 Japanese Requirements

The Japan Highway Research Foundation has published BCO guidelines in its "Design and Execution Manual for Bonded Concrete Overlays" [9] for repairing bridge decks, which is the only commonly accepted design manual for concrete overlays in Japan. It is comprised of six chapters as shown in Table 7.3.

The design principles are presented in chapter 4 of the manual. Mixture proportions for the overlay concrete specify the use of ultra-rapid-hardening cement to get the compressive strength of 24 MPa at the established age (usually 3 hours). This condition gives a higher value for the strength than about 40 MPa when the road is opened for the traffic service.

The minimum thickness required for the overlay is 50 mm, which was determined by taking into account the maximum aggregate size, shrinkage, and the quality of the workmanship. The standard thickness of the treated surface of the substrate concrete is specified as 10 mm. Shot blast methods are recommended for the surface treatment since even shot blasted surfaces can achieve a bond strength of 1 MPa. Although the high performance of the water jet method has been acknowledged in recent studies [10], shot blast methods are still most commonly used in Japan because of their cost. Shear reinforcement is not necessary for overlay concrete. This is based on results of recent research projects related to the interface bond. According to the experimental results of the research projects, a tensile bond strength of 1 MPa between overlay and substrate is sufficient for up to three times higher loads than the design value.

The minimum section area to be overlaid should cover at least the area between expansion joints in the longitudinal direction and both edges for the transverse direction.

7.3.5 Swedish Practice

The Swedish National Road Administration [11], which possesses a long experience on concrete bridge repair, has used and is still using in 2004 the following requirements for BCOs.

The required tensile bond strength is $f_v = 1.0$ MPa. This requirement is satisfied if:

$$m \geq f_v + 1.4 \cdot s \tag{7.10}$$

$$x \geq 0.8 \cdot f_v \tag{7.11}$$

where m and s are the average and the standard deviation ($s \geq 0.36$ MPa) of the measuring values, respectively, and x is a single measuring value.

7.3.6 Expected Design Shear Strength

In this section, the following notation is used:

- Expected design shear bond strength $= f_{b\tau d}$, with f denoting strength, b denoting bond, τ denoting shear, and d denoting design value.
- Expected design tensile bond strength $= f_{btd}$; knowing that the bond tensile strength is about half the bond shear strength, it can be assumed that $f_{btd} \approx f_{b\tau d}$.
- Design concrete compressive strength $= f_{ccd}$.
- Design concrete tensile strength $= f_{ctd}$.

Overlapping with the above strength requirements, different codes or recommendations provide a value that could be considered as a "maximum design shear stress value". It seems that, for carefully constructed overlays, it is the maximum acceptable shear stress calculated with the assumption of a monolithic structure (notably ignoring the built-in stresses deriving from the composite nature of the actual overlaid structure). Because a wide range of behaviour of the actual overlaid structures is ignored, the proposed maximum acceptable shear stresses are very conservative.

According to CEB-FIP MC 90 (1993), the expected design shear strength at the interface is proportional to the compressive strength f_{ccd}. If the surface is rough or indented, the coefficient of proportionality is equal to 0.06. Hence,

$$f_{ccd} = 25 \text{ MPa} \Rightarrow f_{b\tau d} = 1.5 \text{ MPa} \Rightarrow f_{btd} \approx 0.8 \text{ MPa},$$

$$f_{ccd} = 45 \text{ MPa} \Rightarrow f_{b\tau d} = 2.7 \text{ MPa} \Rightarrow f_{btd} \approx 1.4 \text{ MPa}$$

The draft of Eurocode 2 (2001) states that the design shear resistance at the interface is proportional to the design tensile strength f_{ctd} of the weakest concrete. The coefficient of proportionality is 0.45 for rough surfaces and 0.5 for indented surfaces. Hence, at a rough interface,

$$f_{ccd} = 25 \text{ MPa} \Rightarrow f_{ctd} = 2.1 \text{ MPa} \Rightarrow f_{b\tau d} = 0.95 \text{ MPa} \Rightarrow f_{btd} \approx 0.5 \text{ MPa},$$

$$f_{ccd} = 45 \text{ MPa} \Rightarrow f_{ctd} = 3.3 \text{ MPa} \Rightarrow f_{b\tau d} = 1.5 \text{ MPa} \Rightarrow f_{btd} \approx 0.8 \text{ MPa}$$

The American code ACI 318-99 (1999) deals with composite structures in chapter 17. For clean surfaces, free of laitance, and intentionally roughened, the maximum design shear resistance is limited to

$$f_{b\tau d} = 0.55 \text{ MPa} \Rightarrow f_{btd} \approx 0.3 \text{ MPa}$$

The Swedish handbook for concrete structures (1994), for waterjetted properly cleaned interface proposes

$$f_{b\tau d} = 0.4 \text{ MPa} \Rightarrow f_{btd} \approx 0.24 \text{ MPa}$$

None of these values is realistic. Indeed, average measured bond shear strength are in the vicinity of 3 MPa (see Chapter 4 of this volume). The codes of all countries derive the design strength from the measured values through a relationship of the kind of the one presented in Section 7.2.5. With the common scatter of bond strength measurements, it results in $f_{b\tau d} \approx 0.8$ average measured strength. Hence, the realistic expectable value of $f_{b\tau d} \approx 2.4$ MPa $\Rightarrow f_{btd} \approx 1.2$ MPa.

Only the value proposed by CEB-FIP is in the realistic range. All the other proposed values are too low.

In the case of the ACI 318-99 and Swedish proposals, the proposed values are almost one order of magnitude too low. In these cases, it can be supposed that the recommended values correspond, for carefully constructed overlays, to the maximum acceptable shear stress calculated with the assumption of a monolithic structure and ignoring the built-in stresses deriving from the composite nature of the actual overlaid structure. Then, the very conservative values could compensate for ignoring a wide range of behaviour of the actual overlaid structure.

7.4 Conclusion

As already underlined, the existing "design" recommendations or requirements are very limited. Moreover, the few available minimum strength values for the bond are not in agreement. The proposed or required tensile strength ranges from about 0.75 MPa in the USA to 1.5 or 2.0 MPa in Europe. Moreover, major parameters such as, on the one hand the relative thickness of the overlay and the substrate, on the second hand a reinforcement of the overlay, notably by fibres, are not accounted for.

This limited knowledge highlights the need for the development of relevant design methods.

References

1. Barenberg, E.J., Rehabilitation of Concrete Pavement by Using Portland Cement Concrete Overlays, Transportation Research Record 814, Washington, DC, USA, 1981.
2. AASHTO, Guide for Design of Pavement Structures, American Association of State Highway and Transportation Officials, USA, 1993.
3. Army, Army Airfield and Heliport Rigid and Overlay Pavement Design, Department of the Army Technical Manual, TM 5-823-3, USA, 1968.
4. Tayabji, S.D. and Okamoto, P.A., Thickness design of concrete resurfacing, in *Proceedings Third International Conference on Concrete Pavement Design and Rehabilitation*, Purdue University, West Lafayette, Indiana, USA, 1985.
5. Noble, C.S., McCullough, B.F. and Ma, J.C.M., Nomographs for the Design of CRCP Steel Reinforcement, Research Report 177-16, Center for Highway Research, The University of Texas at Austin, USA, August 1979.
6. ACPA Technical Bulletin TB 007-P, American Concrete Pavement Association, Skokie, Illinois, USA, 1990.

7. Felt, E.J., Resurfacing and patching concrete pavement with bonded concrete, in *Highway Research Board, Proceedings*, Volume 35, Washington, DC, USA, 1956.
8. prEN 1504-3, Products and systems for the protection and repair of concrete structures – Structural and non structural repair, European Committee of Standardization (CEN), Brussels, Belgium, 2001.
9. Design and Execution Manual for Bonded Concrete Overlays, Japan Highway Research Foundation, Japan, 1995.
10. Takuwa, I., Shitou, K., Kamihigashi, Y., Nakashima, H. and Yoshida, A., The application of water-jet technology to surface preparation of concrete structure, *Journal of Jet Flow Engineering*, **17**(1), 29–40, 2000.
11. Swedish National Road Administration, General Technical Regulations for Bridges, Publication No. 2004:56, Borlänge, Sweden, 2004 [in Swedish].

Chapter 8
Practice and Quality Assurance

M. Vaysburd, B. Bissonnette and R. Morin

Abstract This chapter discusses construction practice and quality assurance including the importance of substrate surface preparation. The steps in applying a bonded concrete overlay including concrete removal, substrate surface preparation, and application methods for applying the overlay are re described. Methods of quality assurance and construction inspection are discussed.

8.1 Scope and Definitions

This chapter reviews observations of good practices and quality control/quality assurance in the implementation of concrete repair projects involving cement-based overlays.

8.2 Importance of Substrate Surface Preparation

8.2.1 General

The importance of surface preparation cannot be overstated. Proper attention to surface preparation is essential for a durable repair. Regardless of the cost, complexity

M. Vaysburd
Vaycon Consulting, Baltimore, MD, U.S.A.

B. Bissonnette
Centre de Recherche sur les Infrastructures en Béton (CRIB), Université Laval, Québec (QC), Canada

R. Morin
Laboratory of the City of Montreal, Montreal (QC), Canada

and quality of the repair material and application method selected, the care with which deteriorated concrete is removed and concrete and reinforcement surface are prepared will often determine whether a repair project will be successful.

The process of preparation for repair is the process by which sound, clean, and suitably roughened surfaces are produced on concrete substrates. This process includes the removal of unsound and, if necessary, sound concrete and bond inhibiting foreign materials from the concrete and reinforcement surfaces, opening the concrete pore structure, reinforcement damage verification and repair, if necessary.

8.2.2 Safety

Before starting removal of existing concrete, the effect of the removal on the structural integrity has to be reviewed. In cases where removal of deteriorated concrete and/or severely corroded reinforcing steel can affect the load carrying capacity of the structure or its elements, a temporary shoring system should be provided to relive the loads from the structure or its member being repaired. Caution needs to be exercised in order that the safety of the structure is not jeopardized by repair activities.

The areas where concrete to be removed have to be examined to determine if there are electrical conduits, utility lines, or other embedments in the concrete which may be damaged during removal.

All necessary precautions shall be taken to ensure that dust or falling debris do not constitute a hazard to personnel, equipment, the structure, its occupants and the general public. Effective means of clearing dust and debris away from the working area be continuously implemented.

8.3 Concrete Removal

8.3.1 General

Concrete removal on flat horizontal surfaces such as bridge decks, marine piers, etc., differs substantially from concrete removal on other components. The flat horizontal surfaces provide easy access and permit use of high-production mechanized methods. Cleanup and containment of debris are also easier on the horizontal surface.

Other structural components have smaller surfaces that frequently are vertical or overhead and difficult to reach. Difficulties with access and debris containment preclude the use of heavy equipment and make concrete removal less productive and more expensive. The need for scaffolding to provide access for workers and equipment adds to the difficulty and safety hazards involved with the work.

8.3.2 Removal Methods

Concrete removal is addressed in detail in Chapter 4 and in ACI 546R [1].

Removal tasks have to be performed in a manner that ensures that the remaining concrete and reinforcing steel retains its structural integrity. Equipment used to perform the work must not overload the structure, and care must be taken to remove so much concrete that the member being repaired or rehabilitated is critically weakened. Any impact forces used to remove damaged concrete should be applied in a manner that minimizes cracking in the residual concrete and minimizes damage to the bond between the remaining concrete and steel. Methods to remove rust and chlorides from the steel should also minimize damage and loss to the remaining steel.

8.3.3 Advantages and Limitations of Concrete Removal Methods

The study of comparative abilities showed that each concrete removal method had very specific strengths and weaknesses with regard to work characteristics, quality, and production, as follows [2].

8.3.3.1 Basic Methods

1. *Pneumatic breakers.* They are the most flexible and most labor-intensive. They can be used for all sizes and shapes of area, to all depths, and on all bridge structural elements; however, they cause surface microcracking of the substrate, and production rates are low.
2. *Milling machines.* They are the most inflexible. They can only be used to remove large areas of surface and/or cover concrete on decks and are not labor-intensive but capital-intensive. If properly used, they cause little surface microcracking of the substrate and have high production rates.
3. *Hydrodemolition.* This technology lies between pneumatic breakers and milling machines in terms of flexibility. Surface, cover, matrix, and core concrete can be removed; it creates no microcrack and is harmless to the steel reinforcements (bars or welded mesh). Economies are only realized if work is done on large horizontal areas such as decks. Hydrodemolition also has the advantage of selective removal of poor concrete that might be left in place by the breaker method.

8.3.3.2 Combining Methods

Much can be done by combining the strengths and weaknesses of the methods as determined below.

1. *Milling and Breakers*

   Combining these two technologies on surfaces with relatively minor deterioration in cover and matrix concrete results in the following sequence of operations:

   (a) Contaminated and deteriorated concrete is milled to a level conservatively above the reinforcing steel.
   (b) Areas inaccessible to the milling machine are removed using pneumatic breakers.
   (c) Areas of deterioration remaining below the reinforcing steel are determined by sounding and removed to the required depth using pneumatic breakers.
   (d) All exposed steel is sandblasted to remove loose material and rust immediately before repairing damaged areas and overlaying the deck.

2. *Hydrodemolition and Breakers*

   Combining these two technologies on bridges with relatively large areas of deterioration in cover and matrix concrete results in the following sequence of operations:

   (a) The level of contamination and deterioration in the deck is determined using half-cell potential measurement, chloride sampling, and/or sounding. Large areas that require removal are delineated, or a decision is taken to hydrodemolish the whole deck and rely on the selective removal capability of the technique to identify areas of above or below average deterioration.
   (b) Contaminated and deteriorated cover and matrix concrete is removed using hydrodemolition equipment calibrated to achieve the desired results.
   (c) Areas inaccessible to the hydrodemolition equipment and areas of particularly hard concrete are removed using pneumatic breakers.
   (d) All exposed steel is sandblasted to remove loose material and rust immediately prior to repairing damaged areas and overlaying the deck.

3. *Milling, Hydrodemolition and Breakers*

   The efficiency of milling large areas of cover concrete can be used to advantage by using a milling machine to remove all the material above the reinforcing steel. This greatly increases the productivity of the hydrodemolition that follows. The reduction in volume of concrete to be removed by hydrodemolition produces a corresponding reduction in the quantity of wastewater produced, and this greatly improves the environmental and safety impacts. Breakers are used to remove the concrete from areas inaccessible to the milling and hydrodemolition machines.

8.4 Substrate Surface Preparation

8.4.1 General

Overlay surfaces of piers, parking structures, bridges, ramps and loading docks may be exposed to abrasion, rapid temperature changes, salts, moisture, oil, heavy wheel loads, snowplow blades, shrinkage induced shear stresses, live load shear stresses caused by vehicles turning, and impact stresses caused by roughness of the riding surface. The ability of the overlay to survive these conditions is highly dependent on the sound bond of the overlay to the substrate and, therefore, on the adequate surface preparation.

This section addresses the final concrete and reinforcement surface preparation steps taken after the removal of deteriorated concrete necessary to receive the repair materials. The appropriate preparation of the concrete surface depends on the preceding operations and on the type of repair being undertaken.

Concrete removal methods may leave the surface to receive the repair material too smooth, to rough, too irregular, and without open pores. In these cases, procedures specifically intended for final surface cleaning are necessary. Also, microcracking of the concrete surface is common when impact tools are used which may weaken the bond between the existing concrete and the repair. In this case, a less aggressive method of surface preparation such as abrasive or waterblasting is necessary.

8.4.2 Pre-Overlay Repairs

Surface preparation for overlays shall include the pre-overlay repair of defects such as deterioration, corrosion of reinforcement, honeycombed areas, delaminations, small and large holes, sharp protrusions, and cracks. Pre-overlay repair shall consist of all the repair procedures required to bring the substrate to a condition suitable for overlay.

- Damaged sections shall be removed with tools that will not further damage adjacent areas, including the reinforcing steel.
- Careful attention shall be given to the repair of cracks in the concrete substrate.
- The cause of movement in the substrate shall be evaluated and addressed if possible, or the movement shall be accommodated by the overlay.
- The repair materials shall be cement-based and compatible with overlay concrete and substrate concrete being repaired.
- Cure the repairs to be overlaid a minimum of 48 hours or until specified strength has been achieved before beginning final surface preparation procedures. In certain cases, the substrate repairs may be cast monolithically with the overlay material.
- Surface preparation for overlaying shall include the entire surface to be overlaid.

Pre-overlay repairs shall be performed as recommended by ACI 546R [1].

8.4.3 Surface

8.4.3.1 General

Surfaces of existing concrete and exposed reinforcing steel which will be in contact with repair material must be thoroughly cleaned of all bond-inhibiting materials. The concrete surface must be free of microcracking "bruising", dust, and must have an open pore system. All exposed reinforcing steel is to be clean of all grease, dirt, cement mortar, injurious rust, and traces of chlorides from pits in the bars. Injurious rust includes all scale, loose rust deposits, or all rust not firmly bonded to the steel.

When impact tools are used for concrete removal fracturing and microcracking (bruising) of the concrete surface occurs, which weakens the bond between the existing concrete and the repair. Therefore, a less aggressive method of concrete cleaning such as abrasive or waterblasting shall be employed.

8.4.3.2 Concrete Surface Cleaning

(a) General
    First stage cleaning operations shall be commenced in a repair area after all necessary concrete removal has been completed. Concrete surface usually has laitance, partially loosened chips of concrete and bruised concrete layer, removed by blasting. The concrete removal process leaves the resulting surface with varying degrees of microcracks and fractures, commonly referred to as "bruising". Bruising creates a zone of weakness that will affect the bond of the repair material to the substrate. Removal objects the concrete substrate to a wide range of dynamic loads and the resulting bruising will depend on the method used and the quality of the concrete. The depth of the bruised layer varies, but is typically on the order of 3.0 mm.
(b) Available techniques
    Cleaning techniques consist of removing thin layers of surface concrete using abrasive equipment such as sandblasters, shotblasters, or high-pressure water blasters including:

- Sand Blasting – Sand blasting is the most commonly used method of cleaning concrete and reinforcing steel.
- Shotblasting – Shotblasting equipment has the capability to remove finite amounts of sound or unsound concrete. The shot erodes the concrete from the surface. The shot rebounds with the pulverized concrete and is vacuumed into the body of the shotblasting machine.

- High-pressure water blasting (with or without abrasive) – High-pressure water blasting with abrasives is a cleaning system using a stream of water at high pressure with an abrasive such as aluminum oxide or garnet introduced into the stream. This equipment has the capability of removing dirt or other foreign particles as well as concrete laitance thereby exposing the fine aggregate.

8.4.3.3 Reinforcement Cleaning, Inspection and Repair

(a) General
The initial cleaning of exposed reinforcement is usually achieved during the concrete surface cleaning procedures using blasting techniques.
After the initial cleaning, reinforcing steel should be carefully inspected to determine whether the steel shall be simply cleaned or repaired.

(b) Corrosion Protection
When epoxy-coated steel reinforcement is exposed in the repair area it should be recoated with an epoxy coating. Special care must be exercised in recoating operations to achieve defect-free full surface coverage. Uncoated spots may result in severe corrosion in repair areas.
When uncoated reinforcing steel is exposed in the repair area, application of a protective coating should not be done, because it may cause corrosion in areas immediately adjacent to a repair.

A commonly observed phenomenon in concrete repair is increased corrosion activity in existing concrete areas immediately adjacent to a repair. This effect has been referred to as the anodic ring effect, the incipient anode effect, and the halo effect. The repair of corrosion-affected concrete usually addresses the areas where the corrosion activity and related damage is worst – sometimes referred to as the "hot spot". In many cases, it is likely that the conditions in surrounding areas are such that corrosion can occur; however, corrosion activity in these areas has been reduced or dormant due to the active corrosion occurring at the hot spot. This reduction is a form of cathodic protection provided by the hot spot. When the repair is completed, the hot spot has been removed, and the adjacent areas are no longer cathodically protected. In such case the repair area can contribute to corrosion in the adjacent, non-repaired areas. The result is a new "hot spot" which may require additional repair in three to five years.

8.4.4 Moisture Conditioning of the Substrate Prior to Overlay/Repair

8.4.4.1 General

Factors that influence the formation of a bond between repair material and prepared substrate include: the properties of the substrate concrete and its surface, the properties of the repair material, absorption, adhesion, and environmental conditions. Several of these factors are critically dependent on moisture condition of the substrate prior to application of repair materials.

The moisture condition of the substrate will determine the rate of movement of water from the repair mortar to substrate concrete due to the moisture imbalance between the two layers. Both the surface moisture condition and the moisture distribution inside the substrate are important. During the process of water movement two things will occur: penetration of water from the repair mortar in the capillaries of substrate concrete, and hydration of cement paste in the repair.

The optimum moisture condition will vary from substrate to substrate in otherwise equal conditions because the performance of the bond depends on the way the substrate will affect the direction and rate of water movement between phases of the composite repair system.

8.4.4.2 Optimum Water Condition of the Substrate

The optimum water condition of a concrete substrate for a particular cement-based repair material can be determined by preliminary testing using different moisture surface conditioning:

- Saturated Surface Dry (SSD);
- Saturated Surface Wet (SSW);
- Unsaturated Surface Dry (USD); and
- Unsaturated Surface Wet (USW).

In cases when such testing cannot be performed, SSD moisture conditioning should be applied. Under this condition the substrate looks damp but contains no free water on the surface. The surface absorbed all the moisture possible but does not contribute water to the repair material mixture.

8.4.5 Maintenance of the Prepared Substrate

After the substrate has been prepared, it should be maintained in a clean condition and protected from damage until the repair/overlay material is placed.

- In hot climates shade should be provided if practically possible to keep the substrate cool, thereby reducing rapid hydration or hardening of repair material. In wintertime, necessary steps should be taken to provide sufficient insulation and/or heat to prevent the repair area from being covered with snow, ice, or snowmelt water.
- Prepared areas should be protected from repair activities in adjacent areas. Mud, debris, cement, dust, etc., when deposited on a prepared surface, may act as a bond breaker if not cleaned up before the repair material is placed.

8.4.6 Quality Control of Surface Preparation

8.4.6.1 Test Repair

To provide assurance that the surface preparation procedures, materials, installation procedures, and curing will provide the specified repair, the test repair should (it is my understanding that since "shall" is a specification word, we would use "should" instead) be installed at the same thickness and with the same material and techniques, equipment, personnel, timing, sequence of operations, and curing period that will be used for the repair project.

The surface of the designated test repair area shall be cleaned, and the repair material shall be applied, cured, and tested. A test result shall be the average of three tests on a test area of not less than 6 m^2.

8.4.6.2 Required Strength at the Interface

The bond strength shall be tested using the direct tension test. In US practice, it is recommended that bond strength shall be not less than 0.7 MPa. For other countries, the above minimum required tensile strength of the bond and of the substrate may be different (Section 7.2.5). If the failure occurs in the repair material or existing substrate, the bond strength is known to exceed the cohesive strength of the system.

The direct tension test may also be used to determine the soundness of the existing concrete surface before the application of the repair. Tested tensile adhesion strength shall be greater than or equal to 0.7 MPa unless otherwise specified. If the rupture strength is less than 0.7 MPa, the data shall be examined, and the surface preparation procedures shall be modified.

If the strength of the existing substrate concrete is inadequate to achieve the specified repair requirements, the data shall be examined by the engineer and the specified requirements may require re-evaluation.

8.5 Application

8.5.1 General

Experience has repeatedly demonstrated that no step in a repair application can be omitted or carelessly performed without detriment to the durability. Inadequate workmanship, procedures, or materials result in inferior repairs, which will prematurely fail.

8.5.2 Workmanship

It is important that both foremen and workmen be fully instructed concerning procedural details of repairing concrete and reasons for them. Constant vigilance must be exercised to assure maintenance of the necessary standards of workmanship. They should also be apprised of the more critical aspects of repairing concrete

8.5.3 Basic Requirement

The materials used for repairs shall be as recommended by ACI 546.3R [3].

The basic requirement in repair material mixture handling is both quality and uniformity of the material have to be preserved.

Adequate placement capacity should be provided so that material can be kept workable and free of cold joints during the placement. Adequate and consistent supply of repair material during the placement.

The placement process and equipment should be arranged to deliver the material to its final position without segregation. The equipment should be adequately and properly arranged so that placing can proceed without undue delays and manpower should be sufficient to ensure the proper and timely placing, consolidating, finishing and curing of the repair.

8.5.4 Repair Placement

The repair placement should not commence when there is a chance of freezing temperature occurring, unless adequate facilities for cold-weather protection have been provided.

Temperature of the repair material mixture should be between 10 and 29°C, and it should not vary more than approximately 5°C from the temperature of the substrate against which it is placed.

8.5.5 Consolidation

Cementitious repair materials should be consolidated by systematic vibration during placement. Failure to do so, and to do it properly, may result in poor durability, and rapid deterioration. Voids between the repair and substrate can cause total debonding. Voids between embedded reinforcing steel and repair material can cause early corrosion.

The vibration of the repair should be performed by interior vibrators or vibrating screeds or both in combination. However, for relatively small repair areas, application of vibrating screeds is not recommended.

Vibrators should not be used to move repair material laterally. They should be inserted and withdrawn at 15 to 30 degrees from the vertical, so that they quickly penetrate the layer and are withdrawn slowly to remove entrapped air. Vibrators should be placed at close intervals using a systematic pattern to ensure that all repair material is adequately consolidated. The mixture is adequately consolidated when it stops settling, air bubbles no longer emerge, and a smooth layer of mortar appears at the surface.

There is little likelihood of overvibration when the slump of the repair mixture is as low as is practicable for placement using vibration. When overvibration occurs, the surface not only appears very wet, but it actually consists of a layer of mortar containing little coarse aggregate. When overvibration is indicated, the slump, and not the amount of vibration, should be reduced. Efforts to avoid overvibration often result in inadequate vibration. Experience indicates that objectionable results are much more likely to be obtained from undervibration than from overvibration.

Considerably more vibration is sometimes required to satisfactorily reduce the amount of entrapped air and the number of surface bubbles than is necessary to eliminate voids.

Revibration is beneficial rather than detrimental, provided the material is again brought to a plastic condition. Revibration could well be more widely practiced to eliminate settlement cracks and the internal effects of bleeding and also as an aid in making tight concrete repairs.

8.5.6 Surface Finishing

(a) Surface finishing (beyond bull floating) should not be initiated before initial set nor before bleed water has disappeared from the repair material surface. Any finishing of the surface (floating or troweling), while water is rising to the surface will force the bleed water into the repair.

Before initial set, the material is not stiff enough to hold a texture nor stiff enough to support the weight of a finisher or finishing machine. Furthermore, bleeding also controls the timing of finishing operations. Bleed water rises to the surface of freshly cast material because of the settling of the denser solid particles in response to gravity and accumulates on the surface until it evaporates or is removed by the contractor. Bleed water is evident by the sheen on the surface of freshly cast repair material. Finishing the concrete surface before settlement and bleeding has ended can trap the residual bleed water below a densified surface layer, resulting in a weakened zone just below the surface. Finishing before the bleed water fully disappears remixes accumulated bleed water back into the concrete surface, thus increasing w/cm and decreasing strength and durability in this critical near-surface region. Remixing bleed water can also decrease air content at the surface, further reducing durability. Proper finishing should not start until bleeding has ceased and the bleed water has disappeared or has been removed. In most cases, the concrete surface is drying while it is being finished.

(b) When required, the repair should match the existing concrete surface in texture and color.

(c) To obtain a durable surface of the repair, proper procedures should be carefully followed. Following consolidation, the operations of screeding, floating, and troweling should be performed in such a manner that the repair material will be worked and manipulated as little as possible to produce the desired result.
Overmanipulation of the concrete brings excessive fines and water to the surface, which lessens the quality of the finished surface, causing checking, crazing, and dusting. For the same reason, each step in the finishing operation, from the first floating to the final floating or troweling, should be delayed as long as possible while still working toward the desired surface smoothness. Free water is not as likely to appear and accumulate between finishing operations if proper mixture proportions and consistency are used. If free water does accumulate, however, it should be removed by blotting with mats, draining, or pulling off with a loop of hose so that the surface loses its water sheen before the next finishing operation is performed. Under no circumstances should any finishing tool be used in an area before accumulated water has been removed, nor should neat cement or mixtures of sand and cement be worked into the surface to dry such areas.

8.5.7 Curing

Good curing is vital to producing a satisfactory repair. Good curing not only increases durability and wear resistance, but reduces early drying shrinkage cracking. Curing must be started early. Proper curing is even more important for overlays than for ordinary concrete because of the potential for rapid early drying of the relatively thin repair since they have large surface areas in relation to their volumes and moisture can be lost quickly.

8.6 Quality Assurance/Construction Inspection

The following should be considered the minimum inspection requirements to ensure a quality overlay. More than one inspector should be present when the overlay is being placed.

8.6.1 Scarification and Removal of Unsound Concrete

Following the completion of the milling, inspection for delineation and repair of unsound areas should be conducted as described in this chapter.

8.6.2 Substrate Preparation

In assessing substrate preparation, the inspector should determine the following:

(a) Ensure that the surface is free from all latencies, organic residues, and debris.
(b) Ensure that the substrate is in a saturated surface dry condition and free from standing water.
(c) Ensure that the screed rails are properly adjusted to provide the minimum overlay thickness over the entire area to be overlaid, and that the finishing equipment is clean and functioning properly.

8.6.3 Placement and Consolidation

The inspector should account for the following before and during each placement as applicable:

(a) Before placement, a trial batch should be prepared to ensure that the specified slump and air content can be obtained with the job materials and mixture.
(b) Ensure that sufficient amount of material is present to allow continuous placement. Slump tests, total air-content measurements, and a minimum of six compressive-strength specimens should be prepared from each batch. Slump tests should be performed five minutes after mixing.
(c) Monitor the air temperature, material temperature, relative humidity, and wind velocity.
Overlays shall not be placed under the following conditions:

- Air temperature is below 10°C;
- Evaporation rate exceeds 0.49 kg/m^3/hr as determined by ACI 305R.

Overlays shall be placed at night when the daytime air temperature is above 29°C.

(d) Check the levelness requirements of the finished overlay.

(e) Curing

The site inspector should monitor the overlay during the curing period to ensure that it is maintained at a temperature in excess of 7°C and uniformly saturated; membrane-forming curing compounds can be used with double application.

(f) Job acceptance

Following the curing period, the entire overlay should be sounded with a drag chain and hammer to ensure bond. The surface of the overlay should be inspected for cracking. All unsound areas should be repaired by the contractor. Finally, job is accepted or declined based on the results of tensile pull-off tests.

The key requirement of a successful repair is an adequate bond between the repair material and existing concrete substrate, which remains intact throughout its service life. At the present time, practical answers to the problems of bond may depend only on a short-term bond testing rather than long-term performance. An initially achieved adequate bond is only an indication of conformance with the specified parameters. There is no well-defined relationship between initial bond strength and the longevity of a repair. Longevity is influenced by many factors including substrate surface preparation and texture, shrinkage of the repair material, and service conditions.

References

1. American Concrete Institute (ACI), ACI 546R-04, Concrete Repair Guide, 2004.
2. Strategic Highway Research Program Report SHRP-S-360, Washington, DC, 268 pp., 1993.
3. American Concrete Institute (ACI), ACI 546.3R-06, Guide for the Selection of Materials for the Repair of Concrete, 2006.

Chapter 9
Maintenance and Repair of Overlays

D.W. Fowler

Abstract In the event that Bonded Concrete Overlays (BCO) are subjected to distress mechanisms, they may require repair. The types and causes of distress are discussed. Methods for evaluating the extent of distress are discussed, and methods for repairing cracks, spalls and delaminations are described.

9.1 Types and Causes of Distress

Bonded concrete overlays are designed to strengthen and/or rehabilitate existing slabs and pavements and are often a part of the repair process. But overlays sometimes exhibit distress and must be repaired. The most common types of distress are:

1. cracking;
2. delamination; and
3. spalling.

Cracking is caused by one or more of:

1. differential contraction due to thermal stresses or shrinkage;
2. inadequate curing;
3. reflective cracking from substrate;
4. chemical reactions;
5. freezing and thawing;
6. stresses due to external loading; and
7. differential movement of soils or supports.

Cracking often leads to delamination since the stresses due to differential expansion and contraction are the highest at boundaries, e.g. cracks, joints or edges. If the normal or axial stresses at the interface exceed the bond strength between the

D.W. Fowler
The University of Texas at Austin (TX), U.S.A.

substrate and the overlay, the overlay will begin delaminating or debonding at the cracks, joints and/or edges. Delamination is also due to inadequate surface preparation, surface contamination prior to installation of overlays, or premature curing of some bonding agents, e.g. cement slurries or latex slurries.

Spalling can occur at joints due to improper construction or inadequate or non-maintained joint filler. In some cases spalling can occur in the interior areas of overlays due to inadequate curing, expansive aggregate or corrosion of reinforcing.

9.2 Evaluation of Damage

It is important that a condition survey be conducted to determine the extent and causes of damage. The condition survey should:

1. map the location of cracks;
2. mark the location and extent of delamination:

   (a) sounding is the most commonly used technique using a chain drag or a dropped steel rod in a regular pattern on the overlay surface area.
   (b) infrared thermography;
   (c) ground penetrating radar;
   (d) acoustic methods: impact-echo and ultrasonic pulse velocity;
   (e) coring/pull-off tests for determining bond strength; and

3. map the location of spalls and any other distresses.

9.3 Repair Methods

9.3.1 Cracks

If delamination has not occurred at the cracks, the cracks may be repaired by one of several methods:

1. For non-moving cracks, epoxy injection or a topically applied high molecular methacrylate monomer or a low viscosity epoxy resin especially formulated for crack repair can be used. Epoxy injection is quite expensive and is very time consuming, but provides the best structural repair and is usually the best solution if some moisture is present (the cracks should be as dry as possible for the best results.) The topically applied materials are much less costly to apply and are best for extensive cracking in the overlay.
2. For moving cracks which will probably be encountered in bonded concrete overlays only at joints in the substrate where a proper joint was not formed or cut in the overlay or other special cases, an elastomeric joint filler which can accommodate the anticipated joint movement should be used.

9.3.2 Delaminated Concrete

Repairing delaminated overlays requires either rebonding the overlay to the substrate or removing and replacing the overlay in the debonded zone. Rebonding the overlay may be possible using one of several methods:

1. Epoxy injection by drilling access holes through the overlay at regular intervals and injecting epoxy to provide a bonded interface. This repair is usually possible only if the repair is made before the interface becomes contaminated or wet and adequate bonding is compromised. Drying the interface using high pressure air may be helpful if moisture is present.
2. Mechanical bonding using connectors such as steel rods placed and epoxied into drilled holes extending well into the substrate is a possible solution.

Removing and replacing may be the least expensive and most effective repair for many cases. If extensive cracking and delamination has occurred, complete removal and replacement may be required. Prior to repair, however, it is essential to determine the cause of the failure by a thorough forensic analysis to insure that the failure is not repeated. If the complete removal is determined to not be required, the delaminated overlay should be removed beyond the point of delamination by careful saw cutting and careful removal by lightweight pneumatic hammers with care being taken not to damage the remaining sound overlay or substrate concrete. If reinforcing steel or fabric was used in the overlay, it will probably be necessary allow the reinforcing extend into the new concrete overlay a distance to develop the strength of the steel.

The surface preparation, steel placement, concrete mixing and placing, finishing/texturing and curing should follow the procedures for new overlays.

9.3.3 Spalls

The cause of the spalling should be determined prior to repair. The damaged concrete should be carefully removed by chipping hammer or other suitable methods. If corrosion of steel has occurred the concrete should be removed to a depth of at least 20 mm below the bar and the corrosion scale should be completely removed. If expansive aggregate is determined to be the cause the aggregate should be carefully chipped out.

For small repairs, latex-modified mortar or polymer mortar may be the best materials. For larger repairs, latex-modified concrete, polymer concrete or portland cement concrete made with 6 to 10 mm coarse aggregate is preferred.

9.4 Conclusions

Bonded concrete overlays occasionally suffer distress in the form of cracks, spalls and delamination. The extent of distress should first be determined. If the distress requires repair, method of repair that are typically used of other types of concrete distress are suitable for BCOs.

Chapter 10
Conclusion

B. Bissonnette, L. Courard and J.-L. Granju

This State of the Art addresses the relevant issues and challenges in view of achieving durable concrete overlays on a systematical basis.

Various aspects of the current practice have been revisited and validated. Among them, those relating to condition assessment, surface preparation and joint location owe to be reminded.

- In order to select the appropriate repair technique and material, reliable condition assessment is essential. Significant developments in non-destructive techniques recently have helped improving the assessment of the structure to be repaired and establishing a comprehensive diagnosis. Nevertheless, the "old methods" – starting with visual inspection – remain necessary.
- Preparation and cleanliness of the surface to be overlaid is again emphasized to be of paramount importance. Preparation includes the removal of all deteriorated, polluted or corroded material. The removal process must expose sound material with unobstructed porosity, without inducing bruising or microcracking. Then, in order not to jeopardize the quality of the bond, thorough cleaning of the surfaces is required. In most cases, flushing with high-pressure water is satisfactory.
- Locating the overlay joints is also a question of quite significant importance. Most generally, the best approach is still to locate them exactly above the existing joints or discontinuities in the substrate.
- Obviously, attention shall not be paid strictly to the abovementioned steps, as a tight control of all the phases of the repair process is critical. Education and

B. Bissonnette
Centre de Recherche sur les Infrastructures en Béton (CRIB), Université Laval, Québec (QC), Canada

L. Courard
GeMMe – Building Materials, ArGEnCo Department, University of Liège, Belgium

J.-L. Granju
Laboratoire Matériaux et Durabilité des Constructions (LMDC), UPS-INSA, Toulouse, France

training of the workmanship and enforcement of a quality control process are keys to successful overlay applications.

Besides, the State of the Art reviews theoretical and practical considerations relating to bond enhancement, trying along the way to resolve some issues that have been much debated.

- Optimum bond is associated with maximum *effective contact area* (see Section 4.3) at the interface. An overlay material exhibiting a rheological behavior such as to maximize the filling of the smaller asperities of the substrate surface is thus advisable. In this respect, self-compacting overlaying materials can be expected to yield excellent bond.
- The relevance of using of bonding agents is seriously questioned, again. Bonding agents cannot compensate for inadequate surface preparation and may even sometimes act as bond breakers when used improperly. In all cases, grouts as bonding agents should be avoided.
- Placing a rheologically adequate overlay material on a sound and clean SSD (*saturated surface dry*) surface without any bonding agent yields the most consistently successful end results. The expected tensile bond strength then typically ranges within 1.5 and 2.5 MPa, and beyond.

The State of the Art delivers ground-breaking information, notably with respect to the understanding of debonding in overlays and a practical reinforcing approach to prevent it.

- Based on experimental evidence as well as on numerical simulations, it can be concluded that overlay debonding is almost inevitably initiated from a discontinuity, crack, joint or boundary, where the most critical interfacial tensile and shear stresses arise.
- Fiber reinforcement is beneficial in controlling crack opening and propagation in these areas. In addition, it efficiently counteracts the aggravating influence of shrinkage cracking upon the overlay fatigue behavior.
- Fatigue is progressively exhausting the natural interlocking effect in the cracks and along the debonded interface area, thus altering the overall debonding process. At some point, only the interlocking effect provided by the presence of reinforcement (such as fibers), if any, becomes available. On that basis, a conservative calculation to evaluate fatigue effects in fiber reinforced overlays is proposed, allowing only for the reinforcement-induced interlocking.
- The available design rules for overlay bond are rather basic and empirical. In fact, the required minimum bond strength values prescribed in the various consulted codes vary by a factor of up to 3. Moreover, the codes ignore fundamental parameters such as the relative thickness and elastic modulus of the substrate and the overlay, and overlay reinforcement.

More comprehensive design rules taking into consideration the abovementioned issues are needed. Some are currently under development.

The key practical issues highlighted in this State of the Art are at the heart of the final task of this TC's work, which is the publication of *Recommendations* for durable bonded overlays.

RILEM Publications – 9 December 2010

The following list is presenting our global offer, sorted by series.

RILEM BOOKSERIES (Proceedings)

VOL. 1: Design, Production and Placement of Self-Consolidating Concrete (2010) 466 pp., ISBN: 978-90-481-9663-0; e-ISBN: 978-90-481-9664-7, Hardcover; *Ed. K. Khayat and D. Feyes*

For the latest publications in the RILEM Bookseries, please visit http://www.springer.com/series/8781

RILEM PROCEEDINGS

PRO 1: Durability of High Performance Concrete (1994) 266 pp., ISBN: 2-91214-303-9; e-ISBN: 2-35158-012-5; *Ed. H. Sommer*

PRO 2: Chloride Penetration into Concrete (1995) 496 pp., ISBN: 2-912143-00-4; e-ISBN: 2-912143-45-4; *Eds. L.-O. Nilsson and J.-P. Ollivier*

PRO 3: Evaluation and Strengthening of Existing Masonry Structures (1995) 234 pp., ISBN: 2-912143-02-0; e-ISBN: 2-351580-14-1; *Eds. L. Binda and C. Modena*

PRO 4: Concrete: From Material to Structure (1996) 360 pp., ISBN: 2-912143-04-7; e-ISBN: 2-35158-020-6; *Eds. J.-P. Bournazel and Y. Malier*

PRO 5: The Role of Admixtures in High Performance Concrete (1999) 520 pp., ISBN: 2-912143-05-5; e-ISBN: 2-35158-021-4; *Eds. J. G. Cabrera and R. Rivera-Villarreal*

PRO 6: High Performance Fiber Reinforced Cement Composites (HPFRCC 3) (1999) 686 pp., ISBN: 2-912143-06-3; e-ISBN: 2-35158-022-2; *Eds. H. W. Reinhardt and A. E. Naaman*

PRO 7: 1st International RILEM Symposium on Self-Compacting Concrete (1999) 804 pp., ISBN: 2-912143-09-8; e-ISBN: 2-912143-72-1; *Eds. Å. Skarendahl and Ö. Petersson*

PRO 8: International RILEM Symposium on Timber Engineering (1999) 860 pp., ISBN: 2-912143-10-1; e-ISBN: 2-35158-023-0; *Ed. L. Boström*

PRO 9: 2nd International RILEM Symposium on Adhesion between Polymers and Concrete ISAP '99 (1999) 600 pp., ISBN: 2-912143-11-X; e-ISBN: 2-35158-024-9; *Eds. Y. Ohama and M. Puterman*

PRO 10: 3rd International RILEM Symposium on Durability of Building and Construction Sealants (2000) 360 pp., ISBN: 2-912143-13-6; e-ISBN: 2-351580-25-7; *Eds. A. T. Wolf*

PRO 11: 4th International RILEM Conference on Reflective Cracking in Pavements (2000) 549 pp., ISBN: 2-912143-14-4; e-ISBN: 2-35158-026-5; *Eds. A. O. Abd El Halim, D. A. Taylor and El H. H. Mohamed*

PRO 12: International RILEM Workshop on Historic Mortars: Characteristics and Tests (1999) 460 pp., ISBN: 2-912143-15-2; e-ISBN: 2-351580-27-3; *Eds. P. Bartos, C. Groot and J. J. Hughes*

PRO 13: 2nd International RILEM Symposium on Hydration and Setting (1997) 438 pp., ISBN: 2-912143-16-0; e-ISBN: 2-35158-028-1; *Ed. A. Nonat*

PRO 14: Integrated Life-Cycle Design of Materials and Structures (ILCDES 2000) (2000) 550 pp., ISBN: 951-758-408-3; e-ISBN: 2-351580-29-X, ISSN: 0356-9403; *Ed. S. Sarja*

PRO 33: 3rd International Symposium on Self-Compacting Concrete (2003) 1048 pp., ISBN: 2-912143-42-X; e-ISBN: 2-912143-71-3, Soft cover; *Eds. Ó. Wallevik and I. Nielsson*

PRO 34: International RILEM Conference on Microbial Impact on Building Materials (2003) 108 pp., ISBN: 2-912143-43-8; e-ISBN: 2-351580-18-4; *Ed. M. Ribas Silva*

PRO 35: International RILEM TC 186-ISA on Internal Sulfate Attack and Delayed Ettringite Formation (2002) 316 pp., ISBN: 2-912143-44-6; e-ISBN: 2-912143-80-2, Soft cover; *Eds. K. Scrivener and J. Skalny*

PRO 36: International RILEM Symposium on Concrete Science and Engineering – A Tribute to Arnon Bentur (2004) 264 pp., ISBN: 2-912143-46-2; e-ISBN: 2-912143-58-6, Hard back; *Eds. K. Kovler, J. Marchand, S. Mindess and J. Weiss*

PRO 37: 5th International RILEM Conference on Cracking in Pavements – Mitigation, Risk Assessment and Prevention (2004) 740 pp., ISBN: 2-912143-47-0; e-ISBN: 2-912143-76-4, Hard back; *Eds. C. Petit, I. Al-Qadi and A. Millien*

PRO 38: 3rd International RILEM Workshop on Testing and Modelling the Chloride Ingress into Concrete (2002) 462 pp., ISBN: 2-912143-48-9; e-ISBN: 2-912143-57-8, Soft cover; *Eds. C. Andrade and J. Kropp*

PRO 39: 6th International RILEM Symposium on Fibre-Reinforced Concretes (BEFIB 2004), 2 volumes, (2004) 1536 pp., ISBN: 2-912143-51-9 (set); e-ISBN: 2-912143-74-8, Hard back; *Eds. M. Di Prisco, R. Felicetti and G. A. Plizzari*

PRO 40: International RILEM Conference on the Use of Recycled Materials in Buildings and Structures (2004) 1154 pp., ISBN: 2-912143-52-7 (set); e-ISBN: 2-912143-75-6, Soft cover; *Eds. E. Vázquez, Ch. F. Hendriks and G. M. T. Janssen*

PRO 41: RILEM International Symposium on Environment-Conscious Materials and Systems for Sustainable Development (2005) 450 pp., ISBN: 2-912143-55-1; e-ISBN: 2-912143-64-0, Soft cover; *Eds. N. Kashino and Y. Ohama*

PRO 42: SCC'2005 – China: 1st International Symposium on Design, Performance and Use of Self-Consolidating Concrete (2005) 726 pp., ISBN: 2-912143-61-6; e-ISBN: 2-912143-62-4, Hard back; *Eds. Zhiwu Yu, Caijun Shi, Kamal Henri Khayat and Youjun Xie*

PRO 43: International RILEM Workshop on Bonded Concrete Overlays (2004) 114 pp., e-ISBN: 2-912143-83-7; *Eds. J. L. Granju and J. Silfwerbrand*

PRO 44: 2nd International RILEM Workshop on Microbial Impacts on Building Materials (Brazil 2004) (CD11) 90 pp., e-ISBN: 2-912143-84-5; *Ed. M. Ribas Silva*

PRO 45: 2nd International Symposium on Nanotechnology in Construction, Bilbao, Spain (2005) 414 pp., ISBN: 2-912143-87-X; e-ISBN: 2-912143-88-8, Soft cover; *Eds. Peter J. M. Bartos, Yolanda de Miguel and Antonio Porro*

PRO 46: ConcreteLife'06 – International RILEM-JCI Seminar on Concrete Durability and Service Life Planning: Curing, Crack Control, Performance in Harsh Environments (2006) 526 pp., ISBN: 2-912143-89-6; e-ISBN: 2-912143-90-X, Hard back; *Ed. K. Kovler*

PRO 47: International RILEM Workshop on Performance Based Evaluation and Indicators for Concrete Durability (2007) 385 pp., ISBN: 978-2-912143-95-2; e-ISBN: 978-2-912143-96-9, Soft cover; *Eds. V. Baroghel-Bouny, C. Andrade, R. Torrent and K. Scrivener*

PRO 48: 1st International RILEM Symposium on Advances in Concrete through Science and Engineering (2004) 1616 pp., e-ISBN: 2-912143-92-6; *Eds. J. Weiss, K. Kovler, J. Marchand, and S. Mindess*

PRO 49: International RILEM Workshop on High Performance Fiber Reinforced Cementitious Composites in Structural Applications (2006) 598 pp., ISBN: 2-912143-93-4; e-ISBN: 2-912143-94-2, Soft cover; *Eds. G. Fischer and V.C. Li*

PRO 50: 1st International RILEM Symposium on Textile Reinforced Concrete (2006) 418 pp., ISBN: 2-912143-97-7; e-ISBN: 2-351580-08-7, Soft cover; *Eds. Josef Hegger, Wolfgang Brameshuber and Norbert Will*

PRO 51: 2nd International Symposium on Advances in Concrete through Science and Engineering (2006) 462 pp., ISBN: 2-35158-003-6; e-ISBN: 2-35158-002-8, Hard back; *Eds. J. Marchand, B. Bissonnette, R. Gagné, M. Jolin and F. Paradis*

PRO 52: Volume Changes of Hardening Concrete: Testing and Mitigation (2006) 428 pp., ISBN: 2-35158-004-4; e-ISBN: 2-35158-005-2, Soft cover; *Eds. O. M. Jensen, P. Lura and K. Kovler*

PRO 53: High Performance Fiber Reinforced Cement Composites HPFRCC5 (2007) 542 pp., ISBN: 978-2-35158-046-2; e-ISBN: 978-2-35158-089-9, Hard back; *Eds. H. W. Reinhardt and A. E. Naaman*

PRO 54: 5th International RILEM Symposium on Self-Compacting Concrete, 3 Volumes (2007) 1198 pp., ISBN: 978-2-35158-047-9; e-ISBN: 978-2-35158-088-2, Soft cover; *Eds. G. De Schutter and V. Boel*

PRO 55: International RILEM Symposium Photocatalysis, Environment and Construction Materials (2007) 350 pp., ISBN: 978-2-35158-056-1; e-ISBN: 978-2-35158-057-8, Soft cover; *Eds. P. Baglioni and L. Cassar*

PRO56: International RILEM Workshop on Integral Service Life Modelling of Concrete Structures (2007) 458 pp., ISBN 978-2-35158-058-5; e-ISBN: 978-2-35158-090-5, Hard back; *Eds. R. M. Ferreira, J. Gulikers and C. Andrade*

PRO57: RILEM Workshop on Performance of cement-based materials in aggressive aqueous environments (2008) 132 pp., e-ISBN: 978-2-35158-059-2; *Ed. N. De Belie*

PRO58: International RILEM Symposium on Concrete Modelling CONMOD'08 (2008) 847 pp., ISBN: 978-2-35158-060-8; e-ISBN: 978-2-35158-076-9, Soft cover; *Eds. E. Schlangen and G. De Schutter*

PRO 59: International RILEM Conference on On Site Assessment of Concrete, Masonry and Timber Structures SACoMaTiS 2008, 2 volumes (2008) 1232 pp., ISBN: 978-2-35158-061-5 (set); e-ISBN: 978-2-35158-075-2, Hard back; *Eds. L. Binda, M. di Prisco and R. Felicetti*

PRO 60: Seventh RILEM International Symposium (BEFIB 2008) on Fibre Reinforced Concrete: Design and Applications (2008) 1181 pp, ISBN: 978-2-35158-064-6; e-ISBN: 978-2-35158-086-8, Hard back; *Ed. R. Gettu*

PRO 61: 1st International Conference on Microstructure Related Durability of Cementitious Composites (Nanjing), 2 volumes, (2008) 1524 pp., ISBN: 978-2-35158-065-3; e-ISBN: 978-2-35158-084-4; *Eds. W. Sun, K. van Breugel, C. Miao, G. Ye and H. Chen*

PRO 62: NSF/ RILEM Workshop: In-situ Evaluation of Historic Wood and Masonry Structures (2008) 130 pp., e-ISBN: 978-2-35158-068-4; *Eds. B. Kasal, R. Anthony and M. Drdácký*

PRO 63: Concrete in Aggressive Aqueous Environments: Performance, Testing and Modelling, 2 volumes, (2009) 631 pp., ISBN: 978-2-35158-071-4; e-ISBN: 978-2-35158-082-0, Soft cover; *Eds. M. G. Alexander and A. Bertron*

PRO 64: Long Term Performance of Cementitious Barriers and Reinforced Concrete in Nuclear Power Plants and Waste Management – NUCPERF 2009 (2009) 359 pp., ISBN: 978-2-35158-072-1; e-ISBN: 978-2-35158-087-5; *Eds. V. L'Hostis, R. Gens, C. Gallé*

PRO 65: Design Performance and Use of Self-consolidating Concrete, SCC'2009, (2009) 913 pp., ISBN: 978-2-35158-073-8; e-ISBN: 978-2-35158-093-6; *Eds. C. Shi, Z. Yu, K. H. Khayat and P. Yan*

PRO 66: Concrete Durability and Service Life Planning, 2nd International RILEM Workshop, ConcreteLife'09, (2009) 626 pp., ISBN: 978-2-35158-074-5; e-ISBN: 978-2-35158-085-1; *Ed. K. Kovler*

PRO 67: Repairs Mortars for Historic Masonry (2009) 397 pp., e-ISBN: 978-2-35158-083-7; *Ed. C. Groot*

PRO 68: Proceedings of the 3rd International RILEM Symposium on 'Rheology of Cement Suspensions such as Fresh Concrete' (2009) 372 pp., ISBN: 978-2-35158-091-2; e-ISBN: 978-2-35158-092-9; *Eds. O. H. Wallevik, S. Kubens and S. Oesterheld*

PRO 69e: 3rd International PhD Student Workshop on 'Modelling the Durability of Reinforced Concrete' (2009) 122 pp., ISBN: 978-2-35158-095-0; e-ISBN: 978-2-35158-094-3; Eds. R. M. Ferreira, J. Gulikers and C. Andrade

PRO 71: Advances in Civil Engineering Materials, Proceedings of the 'The 50-year Teaching Anniversary of Prof. Sun Wei', (2010) 307 pp., ISBN: 978-2-35158-098-1; e-ISBN: 978-2-35158-099-8; Eds. C. Miao, G. Ye, and H. Chen

PRO 73: 2nd International Conference on 'Waste Engineering and Management - ICWEM 2010' (2010), 894 pp, ISBN: 978-2-35158-102-5; e-ISBN: 978-2-35158-103-2, Eds. J. Zh. Xiao, Y. Zhang, M. S. Cheung and R. Chu

PRO 74: International RILEM Conference on 'Use of Superabsorsorbent Polymers and Other New Addditives in Concrete' (2010) 374 pp., ISBN: 978-2-35158-104-9; e-ISBN: 978-2-35158-105-6; Eds. O.M. Jensen, M.T. Hasholt, and S. Laustsen

PRO 75: International Conference on 'Material Science - 2nd ICTRC - Textile Reinforced Concrete - Theme 1' (2010) 436 pp., ISBN: 978-2-35158-106-3; e-ISBN: 978-2-35158-107-0; Ed. W. Brameshuber

PRO 76: International Conference on 'Material Science - HetMat - Modelling of Heterogeneous Materials - Theme 2' (2010) 255 pp., ISBN: 978-2-35158-108-7; e-ISBN: 978-2-35158-109-4; Ed. W. Brameshuber

PRO 77: International Conference on 'Material Science - AdIPoC - Additions Improving Properties of Concrete - Theme 3' (2010) 459 pp., ISBN: 978-2-35158-110-0; e-ISBN: 978-2-35158-111-7; *Ed. W. Brameshuber*

PRO 78: 2nd Historic Mortars Conference and RILEM TC 203-RHM Final Workshop – HMC2010 (2010) 1416 pp., e-ISBN: 978-2-35158-112-4; *Eds J. Válek, C. Groot, and J. J. Hughes*

RILEM REPORTS

Report 19: Considerations for Use in Managing the Aging of Nuclear Power Plant Concrete Structures (1999) 224 pp., ISBN: 2-912143-07-1; e-ISBN: 2-35158-039-7; *Ed. D. J. Naus*

Report 20: Engineering and Transport Properties of the Interfacial Transition Zone in Cementitious Composites (1999) 396 pp., ISBN: 2-912143-08-X; e-ISBN: 2-35158-040-0; *Eds. M. G. Alexander, G. Arliguie, G. Ballivy, A. Bentur and J. Marchand*

Report 21: Durability of Building Sealants (1999) 450 pp., ISBN: 2-912143-12-8; e-ISBN: 2-35158-041-9; *Ed. A. T. Wolf*

Report 22: Sustainable Raw Materials – Construction and Demolition Waste (2000) 202 pp., ISBN: 2-912143-17-9; e-ISBN: 2-35158-042-7; *Eds. C. F. Hendriks and H. S. Pietersen*

Report 23: Self-Compacting Concrete state-of-the-art report (2001) 166 pp., ISBN: 2-912143-23-3; e-ISBN: 2-912143-59-4, Soft cover; *Eds. Å. Skarendahl and Ö. Petersson*

Report 24: Workability and Rheology of Fresh Concrete: Compendium of Tests (2002) 154 pp., ISBN: 2-912143-32-2; e-ISBN: 2-35158-043-5, Soft cover; *Eds. P. J. M. Bartos, M. Sonebi and A. K. Tamimi*

Report 25: Early Age Cracking in Cementitious Systems (2003) 350 pp., ISBN: 2-912143-33-0; e-ISBN: 2-912143-63-2, Soft cover; *Ed. A. Bentur*

Report 26: Towards Sustainable Roofing (Joint Committee CIB/RILEM) (CD 07), (2001) 28 pp., e-ISBN: 2-912143-65-9; *Eds. Thomas W. Hutchinson and Keith Roberts*

Report 27: Condition Assessment of Roofs (Joint Committee CIB/RILEM) (CD 08), (2003) 12 pp., e-ISBN: 2-912143-66-7

Report 28: Final report of RILEM TC 167-COM 'Characterisation of Old Mortars with Respect to Their Repair' (2007) 192 pp., ISBN: 978-2-912143-56-3; e-ISBN: 978-2-912143-67-9, Soft cover; *Eds. C. Groot, G. Ashall and J. Hughes*

Report 29: Pavement Performance Prediction and Evaluation (PPPE): Interlaboratory Tests (2005) 194 pp., e-ISBN: 2-912143-68-3; *Eds. M. Partl and H. Piber*

Report 30: Final Report of RILEM TC 198-URM 'Use of Recycled Materials' (2005) 74 pp., ISBN: 2-912143-82-9; e-ISBN: 2-912143-69-1 – Soft cover; *Eds. Ch. F. Hendriks, G. M. T. Janssen and E. Vázquez*

Report 31: Final Report of RILEM TC 185-ATC 'Advanced testing of cement-based materials during setting and hardening' (2005) 362 pp., ISBN: 2-912143-81-0; e-ISBN: 2-912143-70-5 – Soft cover; *Eds. H. W. Reinhardt and C. U. Grosse*

Report 32: Probabilistic Assessment of Existing Structures. A JCSS publication (2001) 176 pp., ISBN 2-912143-24-1; e-ISBN: 2-912143-60-8 – Hard back; *Ed. D. Diamantidis*

Report 33: State-of-the-Art Report of RILEM Technical Committee TC 184-IFE 'Industrial Floors' (2006) 158 pp., ISBN 2-35158-006-0; e-ISBN: 2-35158-007-9, Soft cover; *Ed. P. Seidler*

Report 34: Report of RILEM Technical Committee TC 147-FMB 'Fracture mechanics applications to anchorage and bond' Tension of Reinforced Concrete Prisms – Round Robin Analysis and Tests on Bond (2001) 248 pp., e-ISBN 2-912143-91-8; *Eds. L. Elfgren and K. Noghabai*

Report 35: Final Report of RILEM Technical Committee TC 188-CSC 'Casting of Self Compacting Concrete' (2006) 40 pp., ISBN 2-35158-001-X; e-ISBN: 2-912143-98-5 – Soft cover; *Eds. Å. Skarendahl and P. Billberg*

Report 36: State-of-the-Art Report of RILEM Technical Committee TC 201-TRC 'Textile Reinforced Concrete' (2006) 292 pp., ISBN 2-912143-99-3; e-ISBN: 2-35158-000-1, Soft cover; *Ed. W. Brameshuber*

Report 37: State-of-the-Art Report of RILEM Technical Committee TC 192-ECM 'Environment-conscious construction materials and systems' (2007) 88 pp., ISBN: 978-2-35158-053-0; e-ISBN: 2-35158-079-0, Soft cover; *Eds. N. Kashino, D. Van Gemert and K. Imamoto*

Report 38: State-of-the-Art Report of RILEM Technical Committee TC 205-DSC 'Durability of Self-Compacting Concrete' (2007) 204 pp., ISBN: 978-2-35158-048-6; e-ISBN: 2-35158-077-6, Soft cover; *Eds. G. De Schutter and K. Audenaert*

Report 39: Final Report of RILEM Technical Committee TC 187-SOC 'Experimental determination of the stress-crack opening curve for concrete in tension' (2007) 54 pp., ISBN 978-2-35158-049-3; e-ISBN: 978-2-35158-078-3, Soft cover; *Ed. J. Planas*

Report 40: State-of-the-Art Report of RILEM Technical Committee TC 189-NEC 'Non-Destructive Evaluation of the Penetrability and Thickness of the Concrete Cover' (2007) 246 pp., ISBN 978-2-35158-054-7; e-ISBN: 978-2-35158-080-6, Soft cover; *Eds. R. Torrent and L. Fernández Luco*

Report 41: State-of-the-Art Report of RILEM Technical Committee TC 196-ICC 'Internal Curing of Concrete' (2007) 164 pp., ISBN: 978-2-35158-009-7; e-ISBN: 978-2-35158-082-0, Soft cover; *Eds. K. Kovler and O. M. Jensen*

Report 42: 'Acoustic Emission and Related Non-destructive Evaluation Techniques for Crack Detection and Damage Evaluation in Concrete' – Final Report of RILEM Technical Committee 212-ACD (2010) 12 pp., e-ISBN: 978-2-35158-100-1; *Ed. M. Ohtsu*

LaVergne, TN USA
30 March 2011

222127LV00005B/10/P